T0269900

THE HIGH SEAS

OLIVE HEFFERNAN

GREED, POWER AND
THE BATTLE FOR
THE UNCLAIMED OCEAN

THE HIGH SEAS

DAVID SUZUKI INSTITUTE

GREYSTONE BOOKS
Vancouver/Berkeley/London

Imperial measurements have been used in this book for ease of reading, but the original scientific sources cited use metric. In cases where conversion would have been tricky or introduced large errors, we have maintained the original metric.

24 25 26 27 28 5 4 3 2 1

Greystone Books Ltd.
greystonebooks.com

David Suzuki Institute
davidsuzukiinstitute.org

Cataloguing data available from Library and Archives Canada
ISBN 978-1-77164-588-1 (cloth)
ISBN 978-1-77164-589-8 (epub)

Proofreading by Alison Strobel
Indexing by Stephen Ullstrom
Jacket design by Jessica Sullivan and Fiona Siu
Jacket photo by European Space Agency
Text design by Fiona Siu

Printed and bound in Canada on FSC® certified paper at Friesens. The FSC® label means that materials used for the product have been responsibly sourced.

Greystone Books thanks the Canada Council for the Arts, the British Columbia Arts Council, the Province of British Columbia through the Book Publishing Tax Credit and the Government of Canada for supporting our publishing activities.

Canada

Greystone Books gratefully acknowledges the xʷməθkʷəy̓əm (Musqueam), Sḵwx̱wú7mesh (Squamish), and səlilwətaɬ (Tsleil-Waututh) peoples on whose land our Vancouver head office is located.

To Rupie
for everything

To George and Millie
for your future

AND

In memory of Emma O'Kane
I thought our journey would be longer!

CONTENTS

1. The Outer Sea *1*
2. Enter the Twilight Zone *32*
3. The Hunt for Dark Targets *69*
4. Treasures From the Deep *92*
5. The Interventionists *126*
6. A 'Near-Arctic' State *158*
7. The Last Frontier *183*
8. Genes, Drugs and Justice *205*
9. Deep Trouble *228*
10. The Cold Rush *254*
11. Paradise Lost *272*
12. Hope for the High Seas *299*

Acknowledgements *315*
References *322*
Further Reading *346*
Index *347*

1

THE OUTER SEA

FOR MOST, THE high seas are a remote realm, far offshore, that we have neither the chance, nor the desire, to visit. Indeed, the view from an aircraft, cruising at 36,000 feet – a height as great as the ocean is deep – is the closest that many of us come to experiencing this forbidding environment. From this vantage, we can almost grasp the ocean's enormity, its unshifting supremacy in the anatomy of our earthly home. The ocean, after all, occupies 70 per cent of the surface of our planet, and two-thirds of this are 'high seas' – unclaimed waters, beyond national borders. Far from land, these waters are on average nearly 13,000 feet deep, and with this depth comes enormous volume; if we're talking about living space on Earth – the high seas are 95 per cent of what's available. By contrast, all the places that you might visit in your lifetime – the forests, mountains, beaches, deserts and ice caps – comprise a mere sliver of what Earth has to offer. From our lofty position above the clouds, the high seas appear flat, motionless – an almost uniform expanse. Beneath the waves, however, is a world infinitely more complex and varied than the solid ground we tread upon. An immense heaving body of fluid, the ocean is constantly in flux, moving through time and space to connect, heat, enrich and enliven our planet.

From its sun-speckled surface to its lightless depths, the ocean contains places and life forms we have scarcely imagined, many of which are far from shore, and some of which, no doubt, have yet to be discovered.

When I began writing this book, one friend asked what I meant by the high seas. Another asked where to find them. Conversations in the schoolyard at pick-up made me realise that many people have never heard of the most iconic places far offshore. Most people know where the Bermuda Triangle is, and many are now aware of the North Atlantic Garbage Patch, but how many know that the world's only shoreless sea, the Sargasso, is in exactly the same spot? The Sargasso is one of the ocean's most vibrant ecosystems, a place eels will travel thousands of miles, across entire oceans, to reach, so that their young can feast in its plankton-rich waters. Similarly, when I mention Lost City, of course, people think I'm talking about Atlantis – the legendary island that sank beneath the waves, rather than a real place in the Atlantic Ocean. One of the most extreme environments ever found – with lightless, scorching hot, sulphuric waters that teem with bizarre bacteria – Lost City is an area spanning 5,400 square feet that is scattered with hydrothermal vents (much like underwater geysers), one reaching nearly 200 feet tall. Biologists believe Lost City may hold clues to the origins of life on Earth, yet most of us are blissfully unaware of its existence. The same holds true of other places offshore – take the Gakkel Ridge, for instance, an underwater mountain range taller than the Alps, that stretches almost 1,200 miles from Greenland to Siberia, and is home to the eyeless shrimp. Or what of the Saya de Malha Bank, a submerged plateau that formed 120 million years ago when the ancient supercontinent of Gondwana started to break apart, and split into today's major landmasses? In the heart of the Indian Ocean, the waters of the Saya de Malha Bank are shallow and boast seagrass meadows that span 15,500 square miles, making it one of the ocean's largest carbon sinks. Interspersed among these seagrasses are colourful corals and slow-growing encrusting red coralline algae, which attract sea turtles, parrotfish, surgeonfishes and rabbitfishes. Surrounding the bank are deep waters occupied by curious creatures such as the pygmy blue whale and flying fish. Some of the most extraordinary, most biodiverse parts of our planet are on the high seas. Yet they are unknown to most.

Though few of us venture to the high seas, they have long enthralled us. Metaphorically, the high seas, for many, conjure up images of a lawless frontier, a 'wild west' of our planet that harbours outcasts, rogues and opportunists. To others, the open ocean is a place of danger and mystery, for, despite all of our advances in seafaring, these waters remain as inhospitable as they did when early explorers such as James Cook and Christopher Columbus set sail on arduous voyages in search of new lands. Johann Forster, the chief naturalist on Captain Cook's *Resolution*, which sailed deep into the Antarctic high seas in search of an imagined 'southern continent', described the ferocity of a storm that arose on the night of 30 November 1772: 'The people had not yet been prepared for such weather, and therefore did the rolling of the ship much damage; chairs, glasses dishes, plates, cups, saucers, bottles etc. were broken. The Sea came in one or the other cabin and made all the inside wet. In short, the whole ship was a general scene of confusion and desolation.' So tempestuous was the weather on that voyage that Forster looked to the epic Latin poem *Aeneid* for inspiration in describing the scenes he encountered. 'Then came the cries of the men and the groaning of the rigging. Darkness, like night, settled on the sea and all the elements threatened the crew with death at any moment.' Even now, in popular culture the high seas represent a world of storms, shipwrecks and lives lost.

As land dwellers, we've nurtured a fear of this offshore world. Indeed, one of the earliest known maps of Europe – the *Carta Marina*, published in 1539 – shows waters dominated by oversized, mythical beasts, among them the ziphius, a fierce-looking fish that swashbuckles its way throughout the high seas, cutting open its victims and vessels with its sword-like fin; the sea pig, a prickly monster with the feet of a dragon and eyes on the side of its torso, and the Orm, a 200-foot-long sea serpent. Meanwhile, the *Carta Marina*'s landmasses are decorated with representative figures of the time: a sledge pulled by reindeer makes its way through Finland's northernmost territories; a woman fishes by the coast of Norway; a soldier marches on horseback through a volcano-pitted Iceland. But to this day, rare landings of the giant oarfish, which reaches up to 50 feet long and resembles a flattened snake, engender reports of 'terrifying sea beasts' and 'creepy sea creatures'. Our knowledge of marine life has, of course, vastly improved in the past few hundred years: using satellites, we've now mapped 100 per cent of the seafloor to a resolution of 3 miles, allowing us to see major underwater features such as ridges and trenches. Admittedly, we've mapped Mars, the Moon and even Venus in much greater detail, at a resolution of around 300 feet. This is a feat we've not yet achieved for planet Earth owing to its watery shroud. We are now, however, venturing to places we've never been before, illuminating secrets of the high seas. Enabled by a new generation of ships, robots and submersibles, the Seabed 2030 project is aiming to map the entire seabed in high resolution by 2030. This will unveil an entirely new layer of detail and with it, the potential to discover previously unknown worlds. Biologists who study the high seas describe them as 'pelagic waters', which means open sea, far from land. Oceanographers simply call them 'open ocean'. The high seas, on the other hand, is a legal definition – one that describes that part of the ocean beyond national ownership, usually starting 200 nautical miles from shore. And while that may, to some, sound like a curious focus, the simple fact that international waters belong to no one is what makes them so important – roughly half of the surface of our planet, and two-thirds of the ocean, is a vast global commons. It is an unclaimed ocean whose resources we are free to pillage or protect.

MY OWN OCEAN obsession began long ago. From my early childhood, it was there, glimpsed from my bedroom window, if I strained my neck hard enough. The street I grew up on in Dún Laoghaire, a seaside town about 10 miles south of Dublin city centre, ran perpendicular to the harbour and ferry terminal, a location that made it a prime spot for guesthouses, including our own. Ours was a double-windowed, pale green Georgian house with a rose-filled garden. New guests arrived daily, almost always having travelled across seas, with great stories to tell. My most persistent memory, though, is not of their stories or faces but of the bedroom I shared with an elder sister, overlooking the street at the front of our house. Blu-Tacked pictures of rock bands, The Smiths and The Cure, covered its peach-coloured walls, and in their midst was an unsettling image, of two girls looking out to sea, ankle-deep in water, as the ocean rushed in around them, sweeping away their belongings. In the same room, we had a record player and a treasured collection of recorded poems and readings, one of which remains etched in my mind: 'Don't you go too near the sea, for the sea is full of wonder. Don't you go too near the sea, for it's bound to pull you under', boomed the voice, filling my childlike imagination with fear and wonder. Despite years of searching, I've never been able to find evidence of either the image or the recital, as if they too had, somehow, been washed out to sea.

Thus began my fascination with the ocean beyond the horizon. It was a place I longed to visit, to explore and to understand. Eventually, when the time came for university, I chose zoology, with a thesis in marine biology. At weekends, I'd walk to the harbour and watch scuba divers ready themselves before they plunged beneath the waves. I yearned to join them, imagining the subsea world as a place of mystery and beauty, filled with the most extraordinary creatures. Before long, I started to survey local beaches with a non-profit, and was horrified to learn that the waters I'd been paddling in since childhood weren't as pristine as they appeared. Still, it didn't quell my enthusiasm for wading further in. I joined the diving club and saved furiously to buy my first dry suit. I was soon hooked. I spent most of my weekends either on or under water, devoured books about the ocean, and began studying for a PhD in marine ecology, researching the Irish Sea's overexploited fish stocks, including cod, haddock and whiting. By the time I was twenty-three, I had become obsessed with the idea of venturing offshore to the high seas. A fleet of Spanish vessels was fishing for cod just beyond Canada's territorial waters at the time and I was desperate to join them, strangely enticed by the stories of hardship and horror, one from an observer on a trawler whose captain died of heart failure mid-trip and was stored in the freezer with the catch for two months. There were other stories from a trip to the Southern Ocean, less grim, but still unsettling, of months spent at sea with grossly insufficient food supplies, squabbling shipmates and tsunami-like waves that near toppled the boat. When my opportunity came, I grabbed it.

It was July 2001 when I first set sail for the high seas, joining a crew of nine fishermen on a 33-ton fishing boat. That Thursday morning, I boarded a freshly painted red and white trawler, the *Agnes na Mara*,[1] that was to explore a new deep-water fishery for the next three weeks. Our destination was a remote corner of the North Atlantic Ocean, approximately 300 miles northwest of Scotland. We departed from Greencastle, a small fishing village in County Donegal. Tucked into the armpit of the Inishowen Peninsula, on Ireland's north coast, Greencastle was like most other Irish fishing ports back then – quiet, aside from the bustle of the pubs; the harbour filled with brightly coloured trawlers and the docks strewn with fishing gear being loaded on and off the boats.

The boat, and its crew, was part of a fleet of forty vessels from Donegal that had previously fished for cod and other whitefish in the Irish Sea, the same stocks that I was investigating for my PhD. As these inshore stocks diminished, the fishers had struggled to make a living. At the time of my trip, however, government subsidies were covering up to 40 per cent of the cost of new vessels to allow them to travel further to explore and develop an unregulated fishery, one with no quotas, far offshore. There, they targeted unfamiliar species with odd names and even odder appearances: Greenland halibut, orange roughy, grenadier, black scabbard, rabbitfish, and forkbeard (which the Spanish call Sweaty Betty). Some of the catch made its way to fishmongers and restaurants in the UK and Ireland, but most was destined for Spanish and Greek markets. On board *Agnes na Mara*, the crew's

1 Pseudonym. The name of the vessel has been changed.

job was to sound out these relatively unexplored, and untapped, resources. Mine was to note the catch, both intentional and accidental, and to report back to the fisheries authorities in Ireland. In our 33-ton trawler, it took around thirty hours to reach the high seas, where fierce squalls lashed at the windows of our little ship. Days passed when the winds blew too hard and the waves were too high for us to do anything other than ride it out. Despite the weather, we caught plenty.

In the twenty years or more since my trip on *Agnes na Mara*, I've returned to sea numerous times, first as a working scientist, and again since becoming a science writer. Although I've worked in all of the world's major ocean basins, I often reflect on that trip on the *Agnes na Mara*. The crew could not have made me feel more welcome, yet the whole enterprise of deep-water fishing was ill-conceived. A bottom trawler, the *Agnes na Mara* had special heavy equipment that allowed it to fish deep water. On either side of the stern, two wires were attached to a large net held open by steel doors, each weighing around 2.2 tons. Once the net was lowered from the vessel's stern, it was dragged along the seafloor, held in place by the heavy doors, stirring up large plumes of sediment and swooshing fish into the net's opening. The bottom of the net on such a trawler is fitted with large rollers, designed to bounce over rocks and rough ground, protecting the net from wear and tear. Typically, the equipment weighs around 8.8 tons.

It's, therefore, unsurprising that an active trawl fishery, in which hundreds of trawlers are fishing the same grounds, can quickly turn underwater habitats such as corals and seamounts into vast heaps of rubble. The deep sea targeted by these fisheries is also a quiet environment, inhabited by long-lived, slow-growing and late-to-mature species. One of the targets was orange roughy, a species of fish that can live for up to 150 years, and reaches maturity at age 30. Like the old-growth trees in a forest, such individuals are not easily replaced, and their fisheries – with rare exceptions – are unsustainable, often lasting a decade or less. Put simply, we'd moved the problem of overexploitation elsewhere, to a fishery that was less able to recover. Although I didn't realise it at the time, the Irish weren't alone in targeting these high seas stocks. For the Scottish, French, Spanish, Russian, Polish and Faroese vessels all took part in this experimental fishery as part of a much larger global push to discover, and exploit, fishing grounds far offshore. It's now recognised that trawling has caused more severe, widespread and long-term destruction to deep-sea habitats than any other fishing practice, and it's unclear whether these environments can ever fully recover. If they do, it will take centuries, perhaps even millennia.

OUR EARLY ANCESTORS may have navigated deep stretches of water a million years ago, according to a hominid fossil discovered in 2004 on the tiny island of Flores, east of the Indonesian island of Java. This early human species, separate from *Homo erectus*, and dubbed 'the hobbit', may have travelled from as far as Africa to reach Indonesia. By 1000 CE, Polynesian seafarers were crossing thousands of miles of Pacific Ocean to reach neighbouring islands. But it wasn't until much later in human history – around six centuries ago – that people set sail on the first long-distance, recorded, voyages across the high seas. The first to do so were opportunists. As far back as 1402, Emperor Yongle – an ambitious Chinese leader who reigned during the Ming Dynasty – launched a series of seven voyages as far as East Africa and the Persian Gulf under the command of Admiral Zheng He. Each voyage comprised 300 or more 'treasure ships', allegedly nearly 450 feet long, as well as numerous supply ships and warships fitted with canons. In total the crew on each expedition numbered around 28,000. These voyages were intended, in part, to collect treasures from abroad, including foreign spices and jewels as well as strange animals – giraffes, zebras and ostriches – to impress the royal court. Several decades later, at the end of the fifteenth century, Europe's most powerful empires began their own offshore advance. Their ambition to forge ocean trading routes gave rise to the 'Great Age of Discovery', a period recognised for bringing new resources, wealth and knowledge to the west, and for the grave atrocities enacted on the people whose lands were colonised. Routes were discovered from Europe to India and China, before European navigators made their way across the Pacific and pushed south in search of a mythical southern continent, as well as north towards the Arctic. Before long, the high seas became the world's highway; ships filled with precious wares travelled back and forth along well-defined trading routes. European nations such as the Dutch, the Portuguese and the English formed mega-corporations, such as the Dutch East India Company, tasked with ensuring that their wealthy elites had a constant supply of luxury goods. Textiles, silks, coffee and spices all flowed in abundance from the east. These merchant mariners soon attracted a different breed of opportunist: pirates. At the peak of piracy in the seventeenth century, several thousand individuals were terrorising ships on both sides of the Atlantic, targeting vessels leaden with treasures from the New World on their return leg to Europe, and threatening trade between Europe and the Americas. In the Far East, piracy peaked in the early nineteenth century, at which time 40,000 pirates operated a fleet of around 400 junks on the South China Sea, attacking any merchant ship in sight.

These early offshore adventurers – both the merchants and the buccaneers who assailed them – mostly stuck to well-travelled coastal routes, and paid little attention to the natural world. The first to venture off the beaten path were whalers, in pursuit of their prey. As the harpooned giants dived hundreds of feet deep, wrestling with the line, these hunters became unwitting naturalists, sensing the depths beneath them. By contrast, the scientist-naturalists who emerged in the early nineteenth century weren't especially concerned with life on the high seas. Lacking access to the deeper ocean, most assumed that life could only be found at the surface, and that the deep sea was 'azoic' or devoid of life. The 'azoic' theory was conceived of by a British naturalist, Edward Forbes, whose own investigations of the Mediterranean's Aegean Sea led him to believe that marine life became gradually less diverse and smaller with depth, petering out entirely below 300 fathoms (1,800 feet). The first to cast serious doubt on this idea was another British naturalist named Charles Wyville Thomson. In the summer of 1868, he persuaded the Royal Society of London and the English admiralty to lend him a steam frigate, the HMS *Lightning*, to survey the waters between the Faroe Islands and Shetland. There, Wyville Thomson used a dredge to haul up organic remains from the deep sea, evidence that convinced his benefactors to fund further investigation. During three subsequent expeditions to the Northeast Atlantic and the Mediterranean, Wyville Thomson confirmed that life existed at depth.

But it would take the greatest oceanographic voyage of all time to disprove Forbes' azoic theory once and for all. Dissatisfied with the evidence he'd accrued, Wyville Thomson in 1870 approached the Royal Society for further assistance, this time to seek permission to retrofit one of Queen Victoria's ships for a global exploration of the world's oceans. When granted use of a 200-foot-long wooden navy sailing ship, he had all but two of its seventeen guns removed to create space for scientific laboratories and workshops as well as storage for trawls, dredges and samples. Duly refitted, HMS *Challenger* set sail from Portsmouth, England, in 1872 on the first ever oceanic voyage with scientific enquiry as its primary purpose. The expedition covered an astounding 68,000 nautical miles, travelling as far as the Great Ice Barrier of Antarctica, visiting Nova Scotia, the Caribbean and South Africa, before pushing on to explore the Pacific and visiting Indonesia, then heading north to Hawaii and then south again before passing back into the Atlantic through South America's narrow Straits of Magellan. Even now, the scale of *Challenger*'s ambitions seems daunting. Its 243 scientists, officers and crew sampled every one of the Earth's ocean basins. At 362 individual locations, spaced at regular intervals across the seafloor, they took physical, biological and chemical measurements. At these stations and others – 400 locations in total – they used a simple rope marked at regular intervals with flags to take soundings, or depth measurements. The details of the physical measurements and biological specimens they recorded filled fifty volumes and 29,500 pages, all bound in reports that took twenty years to complete.

A few months into the expedition, the scientists hauled up a sea lily, a marine invertebrate that resembles a delicate flower, from water around 6,000 feet deep, just offshore from Lisbon. Closely related to starfish and sea urchins, sea lilies have a central stalk topped with numerous feathered and vibrantly coloured arms, used to catch tiny particles of detritus from the water, especially at night. Later that same week, they collected a Venus flower basket, a strange and magical creature shaped like a conical tube, with walls made of a delicate tissue that resembles spun glass. The heavy nets that *Challenger* used were often lowered as far as 16,500 feet down, and when lifted from the lightless realms beneath proved beyond doubt that the ocean's depths were swarming with unfamiliar species. Among the catch were spectacular and grotesque deep-water fish, never before seen, as well as great hauls of a small silvery fish called the bristlemouth, now thought to be one of the most abundant vertebrates on Earth.

In addition, the scientists made other discoveries, including the presence of ore-rich rocks on the deep seafloor. One especially insightful moment came on 23 March 1875, more than halfway through the expedition. Since leaving Nares Island in Papua New Guinea, HMS *Challenger* had been virtually stranded in the Pacific Ocean. Officer Herbert Swire, the navigator on board, had been assigned the job of taking the depth reading at one of the sampling stations, number 225. With no wind to fill its sails, the ship drifted idly for thirteen days, sometimes moving at only a quarter of a mile an hour. Swire, who kept a detailed journal, noted how the crew's tempers frayed and their faces grew sullen as the days passed with little to occupy them. But on that Tuesday afternoon, as the scientists lowered their sounding rope – a length of simple wire – to the ocean floor, they were astounded. Marked with flags at 25-fathom intervals, their weighted line finally hit the bottom at 4,475 fathoms, or more than 26,000 feet. The incredible gulf below them was by far the deepest point encountered by *Challenger*. Just fifty years earlier, the *Encyclopaedia Britannica* had included the following entry: 'Through want of instruments, the sea beyond a certain depth has been found unfathomable.' No longer. Originally named Swire Deep, Challenger Deep is now known to be part of the Pacific Ocean's Mariana Trench, which reaches down almost 7 miles, or approximately 36,000 feet, and remains the deepest known place in the ocean. Years later, Swire would recall the discovery as his most treasured memory from *Challenger*. Above all, the *Challenger* expedition upended the view of the open ocean as a uniform watery expanse; by its end, it was clear that the open ocean was both deep and filled with life. By that stage, however, and as early as 1870, the first steam-powered fishing vessels had already left European shores for the high seas. With their capacity to fish deeper and to stay offshore for longer came a turning point in our relationship with the ocean offshore – one of unrelenting exploitation that continues to the present day.

IF THE HIGH SEAS are now the wild west of our planet – an unclaimed frontier, open to rampant overuse – one event, in particular, 400 years ago, sealed their fate. It began at 8 a.m. on 25 February 1603, when the crew of a huge Portuguese merchant ship, the *Santa Catarina*, were suddenly woken by a loud crash on the deck. Anchored at the entrance to the Singapore Strait, at the southeastern tip of the Malay Peninsula, the 1,500-ton carrack was heavily laden with goods from China and India, including vast quantities of Ming china, of musk for use in perfumes, and 1,200 bales of silk. The gigantic vessel also carried nearly 1,000 passengers, including soldiers, sailors, women, children and Asian captives whom the Portuguese intended to sell as slaves. That night's anchorage had been a final stopover en route to the Portuguese colony of Macau in southern China, where the *Santa Catarina* would collect yet more riches before sailing back to Portugal with its valuable cargo.

Yet early that Tuesday morning, the *Santa Catarina* found itself under attack by Dutch admiralty sailors who had boarded the carrack on the orders of Captain Jacob van Heemskerck. A merchant seaman, under the employment of the Dutch East India

Company, van Heemskerck had spent months at sea looking for a Portuguese prize. At the time, the Portuguese asserted they had rightful ownership of the eastern Atlantic, a position they'd maintained since 1493, when the Roman Catholic Church had been forced to intervene in a growing dispute between Portugal and Spain over trade routes between Europe and Asia. Drawing an imaginary line from the North to the South Pole, the Vatican granted Portugal the eastern Atlantic with exclusive rights to establish trade routes throughout the region, and Spain the western Atlantic. Eventually, the Dutch began to challenge Portugal's monopoly of the eastern spice route. But the Portuguese fought their ground ferociously. As far as they were concerned, they had the right to defend their territory and, as such, they sought to oust the Dutch from Southeast Asia at every opportunity. In the process many Dutch sailors had been killed, as had locals who granted them port access. One particular incident, involving the execution of seventeen Dutch sailors in Macau in November 1601, had enraged Captain van Heemskerck. Lured ashore by white truce flags, the Dutch had been imprisoned, and then hanged in a Portuguese jail against the wishes of the Chinese authorities. Van Heemskerck had known Admiral Jacob van Neck, the commander executed along with the unfortunate crew, and was devastated by the news of his death. So, when van Heemskerck first laid eyes on the *Santa Catarina*, he saw the ship as a godsend. The attack began at 8 a.m., with van Heemskerck giving his crew orders to fire only at the carrack's mainsails, 'lest we destroy our booty by means of our own cannonades'. With the Portuguese unprepared, the naval battle was fairly one-sided and by six-thirty that evening the *Santa Catarina*'s captain, Sebastian Serraõ, had surrendered, handing over his goods on the condition that the Dutch spare the Portuguese their lives.

The *Santa Catarina*'s capture marked a turning point, in part because her cargo was one of the most valuable seizures on record. When the carrack's wares were later sent to Amsterdam for auction, they fetched the staggering sum of 3 million guilders, today worth roughly $180 million. News soon spread of the riches to be made from trading with Asia, and in particular with China. But the seizure would also go down in history for an entirely different reason. Although the Dutch and Portuguese were at war when the carrack was captured, it wasn't legal under Dutch law for merchant mariners, such as van Heemskerck, to seize goods by force at sea. So the Dutch found themselves in a legally precarious position. To keep the proceeds, they would therefore have to prove that the attack on the *Santa Catarina* was not an act of piracy. As they weighed their options, public interest heightened. Eventually the Dutch called on a young legal scholar called Hugo Grotius.

Grotius was then just twenty-one years old. Drafted in to write the Dutch defence, Grotius marshalled his arguments in a document called *Mare Liberum*, meaning 'the free seas', which was published in 1609. In *Mare Liberum*, Grotius argued that the sea is international territory and that all nations should be free to use it, not just those with existing trade monopolies. In making his case against the Portuguese, Grotius took the moral high ground, arguing that the 1601 murder of Dutch sailors in Macau was not an isolated incident, but one of many barbaric and unprovoked attacks by Portuguese naval powers. In consequence, the Dutch were fully entitled to protect themselves and the *Santa Catarina*'s seizure was a pre-emptive act of self-defence. To bolster his case, Grotius contended that 'Every nation is free to travel to every other nation and to trade with it.' As he wrote:

> *The question at issue is the OUTER SEA, the OCEAN, that expanse of water which antiquity describes as the immense, the infinite, bounded only by the heavens, parent of all things; the ocean which the ancients believed was perpetually supplied with water not only by fountains, rivers, and seas, but by the clouds, and by the very stars of heaven themselves; the ocean which, although surrounding this earth, the home of the human race, with the ebb and flow of its tides, can be neither seized nor inclosed; nay, which rather possesses the earth than is by it possessed.*

Grotius framed *Mare Liberum* as a broader statement of the right to freedom and navigation, sparking an enduring controversy. The ocean was so vast, he contended, that no single nation could appropriate it and secondly, its resources were effectively limitless and therefore inexhaustible. In the long term, the publication of *Mare Liberum* earned Hugo Grotius international renown as a champion of the free seas, and while the idea has faced some stiff challenges, it still forms the basis of how we govern the high seas. In winning his argument, Grotius changed the course of history. For over 400 years, roughly half of our planet has been a vast global commons owned by no one.

Grotius was born in the Dutch city of Delft in 1583, the eldest son of a wealthy family. Regarded as a child prodigy, books became the staple of Grotius' life from a young age. By the age of eight, he was writing Latin poetry; age eleven, he was accepted to study at Leiden University and by the time he was a young teen, he was amending religious texts. Today, Grotius' face adorns the walls of some of world's most prestigious buildings, including the US Capitol, where it sits alongside twenty-two other relief portraits, all depicting the men whose intellectual contributions formed the foundations of modern law. These flattened busts include Thomas Jefferson, who wrote the Declaration of Independence, and Hammurabi, a Babylonian king who crafted one of the earliest surviving legal codes.

London's National Portrait Gallery lists thirteen line drawings of Grotius, created from sittings throughout his life. A painting of Grotius, age forty-eight, adorns a wall in the Rijksmuseum in Amsterdam; another portrait, of him aged just sixteen, can be found in Paris' Frits Lugt Collection, just a stone's throw from the Musée d'Orsay. Arguably the most famous representation, however, is an oversized statue that stands in 'Markt square' in Grotius' hometown of Delft, and which these days is greeted partly with irritation by busy motorists and market-goers.

In addition to these illustrative homages, hundreds of titles detail Grotius' life and works, including at least 350 written during his lifetime. For the most part, accounts of Grotius portray him as a lifelong advocate of peace and of human rights, who was willing to defend his beliefs at almost any cost. Defending the capture of the *Santa Catarina* was just the start of Grotius' career and fame. Sixteen years after the publication of *Mare Liberum*, Grotius wrote *On the Law of War and Peace*, a substantial work in which he set out many of the ideas that form the basis of the laws we use today. Now regarded as the founder of modern international law, he became one of the first scholars to make the case for clemency in conflict. He saw altruism and charity as essential traits of civilised nations, argued emphatically for restraint, and set the tone for how emerging sovereign states should relate to one another. But Grotius' impact on how we manage the high seas has been just as enduring. Who would have thought that a single argument, intended to justify the ransacking of a Portuguese carrack, could – 400 years later – have such a decisive impact on the health of the global ocean? Had Grotius not stepped in to defend the Dutch, would the open ocean have eventually been carved up between nations? And would any nation, over time, have come to protect marine life, to manage fisheries, or to create sanctuaries on the high seas? It is, of course, impossible to say.

At times a great promoter of peace, Grotius was a complicated, conflicted character: he had the propensity to be quite self-serving, even Machiavellian. When Grotius defended the Dutch he was also defending the Dutch East India Company, a conglomerate so aligned with the Dutch state that they were, for all intents and purposes, one and the same. He's been described by one historian as a 'polemic publicist' and an 'itinerant ideologue of empire' whose attachment to the Dutch East India Company was a means to further his own career and that of his family, while also protecting the nation's interests. In the early days, when *Mare Liberum* was being used for the Dutch defence, Grotius was also acting as a political lobbyist and an advocate for the company.

So it would seem that the 'free seas' concept was a smart way of Grotius advancing his own interests, rather than pure altruism. Not long after the publication of *Mare Liberum* – in which Grotius made a passionate plea for the freedom of the seas – he successfully petitioned, lobbied and even threatened the French and English against venturing east to trade in spices and other commodities and was handsomely rewarded with 400 Dutch guilders' worth (US$25,000 in 2023 dollars) of gold coins by the company directors.

For the Portuguese and Spanish authorities, the concept of the free seas proved so controversial that they banned *Mare Liberum* in 1612. The British, too, took issue with *Mare Liberum* and, before long, came up with a rebuttal. Their grievance related to the fact that Dutch vessels were fishing English waters, mostly for herring, and doing so rather more successfully than the English fleet. Throughout the eighteenth century, more nations began to assert their naval powers, leading to conflicting claims on the ocean. To resolve these disputes, in 1702 a 'three-mile limit' was introduced; this gave nations the right to govern waters they could protect within the supposed range of a cannon. For the next 250 years and more, most nations respected what came to be known as the 'cannon-shot rule'. But the peace didn't last. In the 1870s, the first steam-powered fishing vessels were introduced to Europe, allowing fishers to stay at sea for longer, taking more hauls, or to travel further in search of lucrative or untapped fisheries. In the years leading up to the First World War, steam-powered vessels became the norm, and Europe's fishing fleets grew in size and power. This put huge pressure on inshore fish populations, such as North Sea herring and cod, whose stocks started to decline.

By the 1920s, British trawlers started to venture further north to rich fishing grounds around Iceland and the Barents Sea. Soon trawlers from Germany, France, Portugal, the Netherlands and the USSR joined them in fishing the stocks of the far north. Within a few decades, Icelandic fish stocks were also in decline. With Britain and Germany still fishing them hard, in 1958 Iceland called time on foreign fleets and extended their territorial waters to 12 nautical miles from shore. This decision marked the start of the so-called Cod Wars between Britain and Iceland, a bitter dispute that would take almost twenty years to resolve. The British initially ignored Iceland's 12-mile limit, protecting their trawlers with naval frigates, but they conceded in 1961. Then in 1972, Iceland extended its territorial claim again, this time to

50 nautical miles, guarding their waters with patrol boats, and even employing force when deemed necessary.

Eventually, the UN stepped in, and work began to formulate the United Nations Convention on the Law of the Sea (UNCLOS) at a New York conference in 1973. From the outset, it was clear that many nations wanted a 100-mile coastal zone in which they would have exclusive rights over resources. In May 1975, Iceland declared a 200-mile limit. When others followed, the 200-mile limit was written into the Law of the Sea. The treaty was eventually agreed in 1982, came into force in 1994 and has since been ratified by 150 states and the European Union. It allowed states to claim territorial waters or 'closed seas' up to 12 nautical miles from land, and an Exclusive Economic Zone (EEZ), which stretched out to 200 nautical miles, and within which nations had first rights on mineral and fishery resources. Nations could also use these waters to develop wind and wave energy, or to exploit oil and gas, but they remained free for other nations to navigate. There were, however, nuances and exceptions: in certain regions, such as the Mediterranean, EEZS – and, in some cases even territorial seas – clearly overlapped, forcing neighbouring nations to forgo these claims or to negotiate their maritime boundaries.

In fighting for a large Exclusive Economic Zone, nations weren't just interested in fishing rights. Within their EEZ, they also gained first rights to mineral resources – oil, gas and metals – across an area known as the 'continental shelf', which is the submerged portion of a landmass and which, on average, extends out 200 nautical miles from shore. Formed over millions of years, from accumulated layers of organic matter, such as the remains of plants and animals, and inorganic matter, such as sediment from rivers, continental shelves contain around 20 per cent of global oil and 30 per cent of global natural gas reserves, making this submerged territory especially valuable. Some nations, however, have vastly larger continental shelves than others. Along Russia's northern Siberian coast, for instance, the continental shelf stretches almost 8,700 miles from land. Recognising the commercial potential of this subsea real estate, nations were, under UNCLOS, allowed to claim an extension of the 200-nautical-mile allocation, up to a maximum distance of 350 nautical miles from land. First though, they have to prove their rightful ownership using detailed scientific surveys to gather the required data. Predictably, coastal states have jumped at this opportunity, and have collectively, since 2001, asserted their rights to a portion of the subsea twice the size of Russia. The process of deciding who, exactly, owns what is still ongoing.

It took almost a decade to agree the Law of the Sea, a feat that is regarded as one of the greatest ever triumphs in international diplomacy. By the time the deal was done, in 1982, the entire ocean had been carved up into new zones, with fresh rules on ownership and governance. An expanse covering half of our planet was left as a global commons; in a quirk of the diplomacy needed to secure the treaty, the ocean beyond national borders remained the free seas, its resources owned by no one, but the international seabed became a shared commons. Known in legalese as 'the Area', the international seabed is governed, in theory, for the good of humanity, and the profits from its exploitation are shared among nations. But in practice, it's not that clear-cut: the offshore ocean in general is still a frontier, and its unclaimed resources a potential gold mine – whether they are fish from the ocean, minerals from the seabed, or oil and gas reserves from deep beneath the seafloor. Colloquially, experts now refer to all ocean space beyond national jurisdiction as the 'high seas'. In recent years, as scientists, conservationists, lawyers and politicians have wrangled with how to protect the unclaimed ocean, the term 'the high seas' has escaped the confines of the UN, and has entered common parlance. It's with this broader understanding that I use the term throughout this book.

The Law of the Sea has now been ratified by nearly every state on Earth. Even the US, which refused to accede to the convention, abides by its rules. There are huge and undeniable gains that have come from the freedoms it enshrines: it allows us to navigate international waters and the airspace above them safely, it allows nations the freedom to fish and to engage in other activities such as laying cables and constructing artificial islands – if they so wish – offshore. Perhaps above all, it has allowed us to research, explore, understand and benefit from the global ocean. In a world marred by territorial disputes, none of these privileges can be ignored. But in practice, the freedom of the seas has left half of our planet as a free-for-all. That's not, however, to say the high seas are ungoverned: a menagerie of organisations, bodies and codes of conduct oversee specific activities from mining to shipping. No less than seventeen regional fisheries management organisations, for instance, oversee fishing on the high seas, but their mandate is severely limited, covering a tiny number of commercial fish species or certain areas. Vast swaths of the offshore ocean are unregulated for fishing. The International Seabed Authority governs the emerging industry of deep-sea mining, but as we'll see later, part of its mandate is to facilitate exploitation. As a general rule, the high seas have, throughout human history, been managed for commercial gain; preserving the environment offshore has been an afterthought. We have long cultivated an 'out of sight, out of mind' relationship with this half of our planet. And while conservationists have made some recent, serious gains on protecting the high seas, even now industry and economic progress rule the waves. Lax enforcement and apathy remain the status quo. For all intents and purposes, the high seas are still the wild west, a blue frontier.

THE OFFSHORE WORLD has long attracted those seeking fame and fortune. The difference now is that humanity's relationship with the ocean has entered a new phase, marked by the need for new resources. Today's unclaimed ocean is under siege. Far offshore in the Pacific Ocean, for instance, mining companies are exploring the deep seabed for rare metals. As regulators scramble to finalise a rulebook that will green-light this industry, conservationists, scientists and even governments are calling for a ban on bringing mining, a notoriously destructive practice, to the deep sea. Meanwhile, Norway, one of the world's most successful maritime economies, is positioning itself to begin harvesting the fish of the ocean's 'twilight zone', a gloomy mid-water layer between approximately 300 and 3,000 feet that is believed to contain the largest global supply of untapped fishery resources on Earth. Mostly in international waters, the fish stocks they are targeting are unregulated, free for anyone to exploit. Yet the twilight zone also happens to be the ocean's most vital pathway for bringing carbon to the deep sea, where it stays locked up for hundreds if not thousands of years, softening the blow of greenhouse gas pollution.

The overexploitation of coastal fisheries has made a frontier of the high seas, as fleets now need to travel further and to fish for longer in search of ever-dwindling resources. Reaching these rich fishing grounds requires large vessels, plentiful fuel and supplies. As costs are high and it's hard to make a profit, fishers have been known to use unscrupulous means – misreporting the type or the size of their catch, targeting waters where no quotas have been set, or poaching directly from protected areas. Known collectively as 'illegal, unreported and unregulated' (IUU) fishing, these fishing-related offences decimate wild stocks and cost the global economy, through lost taxes, up to $50 billion a year. Throughout the open ocean, environmental transgressions such as poaching are now also associated with more serious crimes such as tax evasion, money laundering, murder, slavery, drugs and human trafficking, as the distant-water fishing industry has become a convenient front for transnational, highly organised criminal gangs. While ferrying drugs from one port to another, for instance, criminals think little of filling their holds with endangered species such as sharks. For the most part, criminals target international waters, outside any nation's jurisdiction, where their violations largely go unpunished. The criminality that now pervades the open ocean is both a symptom and a cause of its overexploitation.

Fresh battles are also emerging offshore, as climate change takes hold. As the Arctic's permanent ice cover melts, China, for example, is planning a shipping route through the newly navigable Central Arctic Ocean, which could eventually become the main passageway between Asia and Europe, supplanting existing sea routes. And the Arctic seabed – currently out of bounds, due to perennial ice cover – could hold up to 13 per cent of the world's undiscovered oil and as much as 30 per cent of the world's undiscovered natural gas. Unsurprisingly, states such as Russia and Canada are now staking claims to the Arctic seafloor, in the hope of extracting these lucrative resources. At the other end of the world, the ongoing conflict between those looking to exploit and protect the high seas is making a political flashpoint of Antarctica's Southern Ocean. Here, shrinking sea ice is squeezing the living space for krill, a tiny brine shrimp that survives by feeding on algae on the underside of ice floes. These partly ice-dependent krill are prey for penguins, fish and whales, and the base of Antarctica's entire food web; they also happen to fuel a billion-dollar industry for nutraceuticals, making them a prime target for the fishing industry.

One could point to the high seas as a modern-day tragedy of the commons, in which the resources of the unclaimed ocean are being exhausted by a few at the expense of all. Take the Pacific bluefin tuna, for instance, a stock that is at around 10 per cent of its pre-fished level, but which saw an increase in its catch quota in 2022, contravening scientific advice. In 2022, the quota for tuna over 66 pounds was increased by 15 per cent in the Western and Central Pacific, compared to 2021. The regional authorities also introduced a new rule, allowing a much larger proportion – 17 per cent – of the previous year's quota to be carried over, compared to 5 per cent originally. And yet this is a stock that is allegedly managed by a sanctioned organisation. For fisheries such as squid, with no catch quotas in international waters, the outcome is worse still: throughout the high seas, squid fisheries are now booming in a regulatory vacuum, leaving experts concerned about the health of the stocks and the larger ocean ecosystem. In the Northwest Indian Ocean, for instance, the presence of squid fishers in one unmanaged area grew by 830 per cent in just five years between 2015 and 2019, threatening the entire local food web.

Added to all this is the problem of pollution. We permit the high seas – our largest earthly commons – to be used as a dumping ground. Space agencies can't dispose of a decommissioned space station or a spent rocket on land or along the coast, but they can do so on the high seas. Farming waste, in the form of excessive nutrients, routinely ends up in rivers and coastal waters, making its way from the coast to the open ocean, where it has helped to create dead zones – vast areas devoid of oxygen and of life – throughout the seas. From 2008 to 2019, the number of dead zones in the ocean increased from 400 to 700, in part a response to nutrient runoff, but also to climate change. Ocean warming is just one manifestation of fossil-fuel pollution; about 30 per cent of our carbon emissions end up in the ocean, changing the temperature, but also altering large-scale circulation and increasing ocean acidity, which has risen 30 per cent since fossil-fuel use began. What's more, so much plastic is now entering the ocean that by 2025, it is expected to weigh 165 million tons.

A general truism of any mistreated commons is that those causing the damage never bear the full cost – hence, their apathy in changing tack. Why would they rejig a system in which they benefit and are rarely held accountable for the damage caused? This is undeniably true of the unclaimed ocean. Flotillas of vessels – armadas even – patrol the high seas squeezing what they can from every inch of unregulated space. Only a handful of nations benefit, but the rest of us suffer the consequences of an ocean that has fewer oxygen-emitting plankton, that soaks up less CO_2 and that provides us with less food for a growing global population. Similarly, those few with an interest in seabed mining stand to profit from this nascent industry, but the damage to the deep sea will be borne by us all.

There are, of course, benefits to our offshore expansion. Since the 1950s, researchers have discovered almost 34,000 marine compounds with commercial potential for a wide variety of uses. An anti-freeze protein from a cold-water fish has been used to improve the texture of ice cream, for instance, and an enzyme extracted from a marine microbe is being used to develop a bio-fuel. So far companies have successfully developed more than a dozen drugs from marine organisms found within national waters. Some have helped solve public health crises and have been worth billions of dollars to the pharmaceutical industry. These include Remdesivir, the first treatment approved for Covid; AZT, the first approved treatment for HIV; and Halaven, a blockbuster anti-cancer drug with annual sales of more than $300 million, all of which have been developed from ocean creatures. Scientists in countries with advanced research programmes are now looking to the unexplored genomes of high-seas organisms for new leads for the marine biotechnology industry, which is projected to be worth $6.4 billion by 2025.

IN JANUARY 2020, a group of researchers from the Stockholm Resilience Centre in Sweden described a new phenomenon, which they named the 'Blue Acceleration'. 'Claiming marine resources and spaces is not new to humanity, but the extent, intensity and diversity of today's aspirations are unprecedented', wrote the authors. The Blue Acceleration is the race between these diverse, growing and often competing interests. The term itself was a nod to the larger phenomenon, known as the 'Great Acceleration', in which our growing consumerism has begun to negatively impact Earth. Since 1950, the human population has trebled in size and has started to use more of the Earth's finite resources, including fish, fossil fuels, wood and water. Since then, we've also physically altered the Earth's surface on an unprecedented scale – by opening mines, building dams and cities, and replacing forests with farms. The changes to our lifestyles have now mapped on to measurable changes in the physical environment. From 1950, there's been an uptick in greenhouse gas levels, global air temperature, ocean acidification, loss of habitat and loss of wildlife, both on land and in the sea.

There are two main reasons for the Great Acceleration's offshore advance. The first is that we're running out of resources on land and so costly endeavours such as seabed mining, once deemed foolhardy, now seem imminent. The second is that technology has advanced to the point where it's now feasible for us to find solutions to complex problems – such as antibiotic resistance – in remote deep-sea environments. As the Blue Acceleration gets underway, the unclaimed ocean is becoming crowded and impacted. Most of the ocean – 59 per cent – is now subjected to cumulative impacts that are becoming increasingly layered. Fishing takes place alongside shipping, military exercises, scientific research, and oil and gas exploration, in seas already awash with plastic and which may eventually be subjected to new pressures such as deep-sea mining. More pernicious still are the warming and ocean acidity that threaten to destabilise the polar ecosystems of the Arctic and Antarctica. The once common view of the ocean – a place of inexhaustible resources, too large to be affected by our actions – has been replaced by a new reality. As the Anthropocene advances, our presence is being felt in parts of the planet that have previously seemed as remote and unknown as other planets.

Faced with a new wave of exploitation, unparalleled in its breadth and potential impact, the story of deep-sea bottom trawling serves as a reminder that our actions have unknown outcomes, some of which may be negative and irreversible. But what if we were to take a different approach, allowing our decisions to be informed by science, and using our global commons in ways that benefit the majority? In writing this book, my aim was to ask whether that is possible. My efforts to explore this question took me to the far corners of the Earth, from the Arctic to Antarctica, and to cities ranging from London to Panama City to Copenhagen. During the course of this journey, I came into contact with conservationists fighting to protect these waters as well as with those looking to exploit the ocean's untapped resources. Above all, I came closer to the huge scientific effort underway to better understand our open ocean, and how we could use this unclaimed space. We might not all agree on the answers. But if half of our planet is a free-for-all, beset by opportunists, who chooses how we use it? And what does that mean for the rest of us?

These decisions are unfolding at a unique point in human history, in which atmospheric greenhouse gas concentrations have reached levels not seen on Earth for 3 million years. Each year, the ocean soaks up roughly a quarter of the emissions that we pump into the air. It's also absorbed 90 per cent of the extra heat we've generated in the past fifty years. To put that in perspective, if all that heat had entered the atmosphere, the global average temperature would now – according to one estimate – be roughly 99°F higher, rendering the Earth uninhabitable. The deep sea, in particular, is our greatest buffer against climate change. The ocean also produces half of the oxygen on Earth, most of which is used to sustain ocean life. But the ocean can only continue as our ally if it remains healthy. A warmer ocean absorbs less carbon, and loses oxygen more rapidly. A stressed ocean, polluted by plastics and overexploited, struggles to support life. To avoid a further collapse of global fisheries and stem the loss of marine biodiversity, scientists say that we need to set aside at least 30 per cent of ocean space in sanctuaries, and manage the rest in a way that is fair and sustainable. As it stands, most marine protected areas are ineffective, blighted by a lack of regulation and poor oversight. Just 1 per cent of the high seas is fully protected.

Yet, there are reasons to be hopeful. We are entering an era of unrivalled political will to protect the high seas. In March 2023, UN member states secured an historic deal that could allow large swaths of the high seas to be designated as marine protected areas, and that could slow our offshore advance. It's now conceivable that places such as Lost City or Gakkel Ridge or the Sargasso Sea will be safeguarded for future generations. What's more, we are witnessing an era of rapidly evolving ocean surveillance technology, bringing visibility to the offshore world and accountability to the industries that operate there. Our scientific understanding of the ocean has never been greater. Yet despite these wins, even now, the offshore world remains an untamed frontier. To redress our 'out of sight, out of mind' relationship with the unclaimed ocean, and to use our ocean commons in ways that benefit the majority, we need to go further. In the years that I've been writing about this topic, we've slowly come to accept that if we don't change tack, we risk doing irreversible harm. My hope, in writing this book, is that everyone, and not just ocean obsessives like me, will understand what's at stake.

2

ENTER THE TWILIGHT ZONE

ON 21 SEPTEMBER 2021, an enormous, odd-looking creature was projected in purple onto the northern façade of the UN building in New York City. Some thirty storeys tall, the image was the creation of Superflex, a Danish art collective, and the animal they featured was a siphonophore. Dancing 500 feet above the ground, its wiry tentacles stretching out in every direction, the siphonophore was there to remind passers-by of the unsung creatures of the ocean's twilight zone, of which it is one. A year earlier, scientists in Australia had discovered a siphonophore almost 430 feet long, believed to be the longest existing animal on Earth. A relative of the jellyfish, a siphonophore is actually a colony of gelatinous animals called 'zooids' which, attached together, form long transparent floaters that drift with the currents. They inhabit a mid-layer of the ocean, where temperatures drop as low as 39 °F, visibility is poor, and much of the available food rains down from the sunlit waters above.

If ever there was a hang-out for weird creatures, it's here, in the twilight zone. Mostly far offshore, this ocean mid-layer starts where photosynthesis stops, at around 300 feet deep, and ends where the ocean becomes truly devoid of light, at roughly 3,000 feet. Though bizarre, the siphonophore is by no means the oddest of the twilight zone's inhabitants: competitors for that title could be the Sloane's viperfish, an agile swimmer no bigger than a human hand. With fangs too large to fit inside its mouth, the Sloane's viperfish impales its victims in high-speed attacks, which are so forceful that it's adapted its first vertebra to withstand the impact. Then there's the dragonfish, a scaleless species that uses red spotlamps to illuminate its prey and a barbel, dangling from its chin, to lure them. Or the bristlemouth – a small fish, of thirty-one separate species, all with the ability to glow in the dark and most with huge gaping mouths filled with sharp needle-like fangs. Thankfully, the bristlemouth, which measures just 1–3 inches long, poses only a threat to tiny fish, shrimp and plankton.

In the ocean, siphonophores writhe and sway with the currents, performing elaborate, if unintentional, dances. They perform other, more important, rituals too, migrating from their home in the deep, where they hide during daylight, to feed at the surface at night in what is now recognised as the largest animal migration on Earth. This daily undertaking involves thousands of species, and trillions of individual animals, and through a variety of means it transfers carbon, in bulk, from the Earth's atmosphere to the deep sea. Superflex's *Vertical Migration* installation – projected during UN Climate Week 2021 – also served as a reminder that, every day, siphonophores and their mid-water allies provide our atmosphere with a vital carbon-removal service. The scale of this operation is immense. If you 'sucked out all of the twilight zone's organisms – you would more than double the amount of CO_2 in the atmosphere that humans have added', said Ken Buesseler, an ocean biogeochemist. 'That would take us to temperatures we haven't seen in 20–30 million years, when the Earth was 12° Celsius warmer than today, there was no ice on either pole, the seas were much higher, and many parts of the world where people now live were uninhabitable.' Far from being the oceanic desert imagined by late-nineteenth-century biologists, such as those on the HMS *Challenger* expedition, the twilight zone is now believed to be the largest untapped supply of fish on Earth. Conservative estimates now put the total mass of fish here at just over a billion tons. But the volume of fish in the twilight zone could be exponentially larger; a 2014 study suggests there could be 10 billion tons or more of fish here – an extraordinary figure which means that, if true, the ocean's twilight zone contains a staggering 95 per cent of all of the ocean's fish if measured by weight alone.

Although the 10 billion figure is hotly disputed, the implication is immense. An unregulated resource, this zone is mostly beyond national waters and free for anyone to exploit. A valuable larder that we haven't yet raided? A new fishery for any nation that could make this work? Accompanying the 2021 installation by Superflex in New York was a panel discussion by four ocean explorers, one of whom was film director James Cameron. 'The twilight zone is full with trillions and trillions of animals. On any given day there are more of them down there than there are stars in our galaxy', said Cameron. 'And yet commercial fisheries are getting really interested in how we can use this biomass and how can we capture it, how can we harvest it, how can we feed it to fish and sell it to people. Typically whenever we humans get interested in something economically we start destroying it. It's just a pattern.'

In recent years, Norway has positioned itself as the front-runner among nations interested in fishing this resource. More than any other nation, Norway has thrown money and research time at developing an experimental fishery for 'mesopelagics', the industry term for fish that inhabit the ocean's twilight zone or 'middle layer'. Their quarry are the lanternfish and pearlsides, each roughly a quarter of the size of a sardine, which inhabit Norway's Exclusive Economic Zone (EEZ), but are mostly found in international waters. If Norway is successful, this will be the first high-seas fishery the maritime economy has developed in over thirty years. By the time of the Superflex show, I'd been following the conversation about the twilight zone, and the fate of its inhabitants, for some time. I'd followed Norway's sea trials closely, but from afar. Curious to find out more, a couple of months after the Superflex siphonophore show, I headed to Norway.

THE SINTEF SEALAB, a grey, fairly nondescript building, occupies a prominent position on the dockside in Trondheim, on Norway's west coast. An oceans-focused branch of one of Europe's largest independent research organisations, the SINTEF SeaLab is involved in all sorts of ambitious and interesting projects – developing the world's first electric fishing boat, for instance, inventing biodegradable nets that will dissolve if lost at sea, and, over the past few years, exploring the untapped commercial potential of the twilight zone. Since 2016, the SINTEF SeaLab has been coordinating Norway's experimental fishery for mesopelagics in national and international waters. The research programme, which is co-funded by the government and by industry, has involved forty vessels. Some private companies have invested heavily in this effort, with one, named Birkeland AS, having spent over US$4 million – a remarkable sum for an experimental fishery – and 120 days at sea by 2021. These fishers typically catch mackerel, herring and blue whiting, a relative of cod usually processed into fishmeal and oil. Now they are targeting the small silvery Mueller's pearlside (*Maurolicus muelleri*) and the larger goggle-eyed glacier lanternfish (*Benthosema glaciale*), among the twilight zone's most potentially lucrative fisheries. Since 2020, the Norwegian government has also funded a new research centre at the SeaLab to help develop this fishery. Scientists here are investigating how fish from the twilight zone could be caught, harvested, processed and turned into various products – mostly fish oils and fish feed – to grow Norway's booming salmon farming industry. But the first, formidable, challenge is catching them.

I arrived at the SeaLab on a Monday in November of 2021. The outside temperature was around 39 °F, the air damp from a light but constant drizzle. To reach the dockside from town, I made my way past the harbour – with its day cruisers, small fishing boats and charming colourful warehouses – over a wide bridge, and past an array of high-rise offices and modern hotels before arriving at the SeaLab. The timing couldn't have been better. In the months leading up to my visit, Norway had come under increasing attack for its heavy exploitation of international mackerel, herring and blue whiting stocks. These fish stocks are transboundary, meaning they move between national waters and the high seas; they are also what's known as 'straddling stocks', in that they move between the EEZs or territorial waters of several nations. The point is that their fisheries should be managed jointly by the nations that exploit them. But for more than a decade, the six states that target these stocks – the EU, Iceland, the UK, Russia, the Faroes and Norway – have set their own individual quotas, in effect disregarding the catch limits advised by scientists. Since 2015, the combined catch of these stocks has exceeded the recommended limit by almost 5.5 million tons a year, and in 2020, all three – mackerel, herring and blue whiting

– were stripped of their certification as sustainably managed fisheries. Undeterred by the backlash, Norway unilaterally increased its own mackerel quota by 55 per cent in 2021, a move the European press labelled as a 'mackerel grab'. Tensions had risen to the extent that by July 2021, for instance, the Irish port of Killybegs in Donegal was refusing to allow Norwegian vessels to land their catch. Rupert Howes, chief executive of the Marine Stewardship Council, a body that awards certification to fisheries, said: 'The most worrying thing is the stocks are trending down ... there is a growing voice from the market, which is saying this needs to be resolved.' The week before my visit to Trondheim, the six nations involved had gathered in London to resolve the dispute, but again failed to agree on a fair split.

With Norway so clearly overfishing the high seas, was it sound to start a new fishery, I wondered. My contacts at the SINTEF SeaLab were many but, by the time of my visit, I had narrowed it down to a couple of scientists who were willing, at least, to discuss Norway's plans. They were Eduardo Grimaldo, a slight, muscular Peruvian, whose job was to develop the gear that would catch the fish, and Dag Standal, a tall, burly Norwegian focused on how this fishery might be governed once underway. In a conference room in the SeaLab, overlooking the dock, with its own assemblage of pretty boats, Dag took a marker and started drawing a simple graph with two lines on a whiteboard: one line represented the ageing Norwegian population, and the other the projected decline in revenue from Norway's oil and gas industry. This was the problem that Norway, currently one of the world's wealthiest petrostates, needed to solve. Without serious changes to their economic model, he explained, Norway could be facing a serious fiscal crisis in as little as thirty years. 'The big challenge in Norway today, and for the twenty to thirty years to come, is to fill this gap', said Dag, 'and as an ocean-based economy, we're looking at marine industries.' When, in 2012, SINTEF proposed that salmon farming be expanded five-fold by 2050, the rationale was straightforward: the Norwegians are already so good at it. The country would simply scale up from producing 1.4 million tons of farmed salmon each year, to 5.5 million tons annually. This expansion would mostly serve overseas markets such as Europe and the US, but also China, Korea and Japan, and the government liked the idea enough to adopt it as a national strategy. The scale of this challenge is immense, not least because of the roughly 6.6 million tons of fish feed required each year to grow 5.5 million tons of salmon. The many possible solutions include rearing soldier flies en masse in factories to feed to salmon, or growing feed such as yeast using single cell technologies and farming algae. Another option is catching fish from the ocean's twilight zone and turning it into fishmeal.

Norwegian salmon feed is currently made from about eight or nine ingredients, shipped to Norway from all over the world, including food colouring from China, beans from Ukraine, wheat from Europe, krill from the Southern Ocean, and soy from a variety of places including Brazil, where it is grown on land that was once virgin rainforest. A mere 10 per cent of the ingredients in Norway's fish feed today are sourced locally, a big shift from thirty years ago, when most feed was made from wild-caught fish. Now, a quarter comes from the sea; this includes krill, fish oils and cast-offs or 'trimmings' from the fishing industry. The rest is derived from land-based sources, and most of that is soy, about 60 per cent of which comes from the tropics. Once the salmon has been fed, farmed and processed, it too is shipped all over the world, sometimes using a supply chain that defies all logic: salmon farmed in Norway can be sent to France for filleting and packaging, only to be shipped back to supermarkets in Norway and sold as fresh fish, or shipped onwards to other destinations such as the US and the Asian market in China, Korea and Japan. This is an industry with an outsized carbon footprint, a problem with which conscientious consumers are becoming increasingly impatient.

The reasons for fishing the twilight zone are that it could help expand aquaculture, make fish feed that is more nutritious (with more omega-3s), is sourced by Norwegians and has lower emissions, by replacing the need for soy products that come all the way from Brazil. Norway's hope is to harvest around one-tenth of the fish – by weight – available in the ocean's twilight zone each year, which by some calculations, is just over 1 million tons of commercial catch. That's one-sixth of the feed needed for the salmon industry, or about three times what Norway now catches in mackerel, by weight. It's an ambitious aim, but if Norway succeeds, it would justify, economically, the research efforts that are currently underway.

There are of course, problems with harvesting this amount of fish from the twilight zone, not least due to the fact that in the wild, these small 'mesopelagic' fish are nature's carbon capture devices: by feeding on carbon-rich foods in the ocean's shallow waters and then migrating downwards, where they excrete and occasionally are eaten, they transfer carbon, in bulk, to the ocean's depths, where it stays locked out of the atmosphere. These mesopelagic fish are also the food source for internationally important wild fish stocks such as bigeye and yellowfin tuna, and for the mackerel and herring stocks that are currently being fought over, and depleted, by Norway and other nations; fishing their food will – arguably – add another layer of stress to the system. Similar conflicts are currently playing out in Norway's own waters: with the help of SINTEF's scientists, in 2019 Norway opened a new fishery for *Calanus*, a genus of tiny zooplankton that are the food of twilight zone fish, but also of larger species such as cod; predictably, this has caused a rift between those fishing cod and those fishing *Calanus*.

With Norway planning to start yet another fishery – this time for mesopelagics – Eduardo's job is to help fishers catch them. Around twenty years ago, before joining SINTEF, Eduardo worked designing nets and other gear in his home country of Peru, a nation famed for its anchovy fisheries. At that stage, Peru was also keen to fish for mesopelagics in its offshore waters; lanternfish were turning up in large shoals off the coast of South America, and so scientists were asked to assess the size of the stock, and to find a way to catch the fish. Despite finding large, if variable, quantities, the government decided against starting a commercial harvest for fear it would take food away from their lucrative anchovies. 'And then the case was closed', said Eduardo. 'Nobody is bothering any more.' After that, Eduardo was recruited by SINTEF to solve the same problem in Norway: could he help them to find, and catch, fish in the twilight zone? Relocating to Trondheim in 2016, he has since been trying to find these fish in commercial quantities, an effort that, he told me, has taken him far and wide across the Atlantic, from Southern Iceland to the Azores and west of Ireland.

HIGH-SEAS FISHING HAS existed ever since Europeans first ventured as far as Newfoundland in the sixteenth century in search of cod. The scale and reach of high-seas fishing, however, was revolutionised with the invention of the combustion engine in the late nineteenth century. By the 1920s, as inshore fisheries for cod, flatfish and other species had started to decline, British and European trawlers started to move further offshore, making their way further north to rich fishing grounds around Iceland and the Barents Sea. The same trend played out off the coasts of New England in the US, along the coast of Japan, in China and throughout Asia, everywhere in fact that larger steam-powered fishing vessels moved out into the world's open waters.

But during the Second World War, high-seas fishing really took off. Engines suddenly became more powerful, refrigeration units improved and sonar developed to the point where it could be used to track large marine mammals or schools of fish, challenging fishers to try their luck further from home. Since then, fisheries have expanded offshore globally, first in the northern hemisphere and then south. Since the 1980s, the southward expansion has progressed most rapidly, at a rate of almost one degree of latitude per year. As fishers have ventured further from home, they have also fished deeper, as far down as 5,000 feet in the northeast Atlantic, for instance. By 2010, fishers were targeting waters 1,150 feet deeper, on average, than they had been in 1950. By the mid 1990s, very few parts of the high seas – only the unproductive regions, and ice-covered waters of the poles – remained as frontiers for the fishing industry. While it's unclear how much of the ocean is now being fished – estimates vary wildly from 4 per cent to 55 per cent of the surface area – it has had a sizeable impact on the high seas, where fish stocks are often unregulated and where enforcement is lax. As fishers have voyaged further afield, often backed by government subsidies to cover their costs, they have targeted their catch with escalating vigour.

These days, there are few opportunities to develop new offshore fisheries. If Norway succeeds in developing a commercial twilight zone fishery, it will not be the first nation to have tried.

One of the earliest attempts was by South Africa in the 1980s, at a time when the country was starting to worry about the depletion of its coastal fisheries. With evidence of a vast, mid-water population of Hector's lanternfish (*Lampanyctodes hectoris*) – a small fish of the twilight zone, reaching around 3 inches long – just off the African continental shelf, estimated at 20 million tons, the government decided to fund fishing trials in the hope of securing a new source of revenue. But mesopelagic fisheries have their challenges – the first is catching small fish that don't form dense shoals, but rather move as isolated individuals through a vast ocean. Added to that are the difficulties in processing fish with a high oil content, making them nutritious but also highly flammable. In South Africa, the risks posed by leaving oily, and combustible, lanternfish in the sweltering heat soon became evident when the catch caught fire, engulfing an onshore processing plant, bringing a swift end to the entire venture. Elsewhere, noxious substances, including cadmium and arsenic, have been found at higher-than-desirable concentrations in mesopelagic fish. Lanternfish from the Norwegian fjords and from the Northeast Atlantic, for instance, have relatively high concentrations of wax esters, fatty substances that can cause severe diarrhoea and even a skin rash called seborrhoeic dermatitis in humans.

None of this, however, has deterred those keen to start a new fishery. Following South Africa's efforts, Russia, India, Oman and Iran have all attempted to harvest the twilight zone off their own coasts, but these efforts have been small and hindered by high costs and low catches. Iceland has been somewhat more successful. In 2009 and 2010, their attempts to catch pearlsides (another potentially commercial twilight zone species, usually smaller than lanternfish, at around 1.5 inches long) in their own waters went well enough for them to seek expert advice on establishing a quota. But as soon as they set the limit – 33,000 tons for 2013 – the fish vanished. In December 2009, Icelandic vessels had caught 54,000 tons, with catch rates of up to 615 tons per day. In 2014, the catch was zero. 'Iceland had a big fishery, some good resources and then suddenly the fish disappeared. It was there and then it disappeared', Eduardo told me. During fishing trials between 2014 and 2017, the Spanish identified a potential fishery in the Bay of Biscay but, like the Peruvians, concerns over taking food eaten by their usual targets – in this case mackerel – made them hold off on developing it further. In 2016, the US went further still, when its Pacific Fisheries Management Council placed a moratorium on the harvest of mesopelagic fish in EEZ waters off the US west coast. Conservationists, and some scientists, are now calling for this ban to be adopted worldwide, at least until the potential impacts are fully assessed.

At SINTEF, Eduardo and Dag support a precautionary approach – a fishery can't and won't be developed, they say, until they know whether it's sustainable. But that information could take at least five years to gather, they told me, all while Norway is keen to start expanding its aquaculture industry. 'When we start to exploit a new fishery which is unregulated, that is called positioning', Dag said. For the moment, Norway is positioning itself. It's the same approach that Norway and others, including Europe and Britain, used to establish their right to fish for blue whiting decades ago. The fishery began in the 1970s and developed throughout the 1980s; in 1987 the catch was 806,140 tons. 'It was unregulated at that point', explained Dag. 'There was no quota in national or international waters.' Nations couldn't agree on where the stock, which is transboundary – moving around between the high seas and national waters – was positioned and how much each nation should get. The result was rampant overfishing: by 2004, the catch was 2.65 million tons, forcing the regulators to step in, close the fishery and grant a limited number of licences. 'The vessels that had already been out there had gained an historical right. They could show their record of fishing these grounds.' If a fishery starts in the ocean's twilight zone, it may well start as a free-for-all, without access restrictions or quotas. If it later becomes regulated, the Norwegians, having been there first, will have first rights. 'These forty companies in Norway, who applied and got these permits', said Eduardo, 'they did this to position themselves, to get this historical advantage and to be first in the queue.'

ON LEAVING THE SINTEF SeaLab, I made my way to the Trondheim office of BioMar, one of the world's largest suppliers of fish feed, just a five-minute walk away on the dockside. I was there to see Vidar Gundersen, who leads BioMar's global sustainability programme. Ascending in a glass elevator, we entered a warm office looking out over the Norwegian fjords. As a company that gets 70 per cent of its business from the salmon aquaculture industry, I wanted to know what they thought of Norway's salmon expansion plan, whether it was feasible and whether they envisaged a future where twilight zone fish were being used in their feeds. But I also had another reason for paying them a visit: the day before, on exiting the airport at Trondheim, I had clocked a huge billboard emblazoned with BioMar's logo that read 'Aquaculture is Norway's largest renewable export industry'. Given everything I knew about the industry, I was doubtful of its claim to being renewable, and I wanted to know more.

Vidar explained that BioMar are constantly striving to make fish feed more sustainable. But it's tricky: as they improve on some aspects of sustainability – such as water use or land use

– others can worsen. Over the past decade the amount of wild-caught fish in the feed has decreased, for instance, and has been replaced with soy, growing its carbon footprint and reducing its omega-3 content, the major selling point of salmon. The struggle now is to lower the product's associated emissions and increase the nutrition. My mind boggled as I struggled to imagine scaling the fish-feed industry up – sustainably – five times over by 2050. On top of that, unexpected complications can arise. Vidar pulled up a slide explaining that in the 1990s bloodmeal was used as a source of animal protein in fish feed, but after the Mad Cow Disease outbreak they couldn't be sure if it was safe. So they needed to find alternatives. I wondered aloud where mesopelagics would fit in especially at the scale imagined by SINTEF, with Norway harvesting just over 1 million tons a year of the stuff. 'It's a vulnerable ecosystem, and whether we use it – I think this will be a debate for a very long time', said Vidar. He then offered to introduce me to his colleague Roger Elvestad, who sources marine ingredients for BioMar and who would ultimately decide whether fish from the twilight zone would become a component of their fish feed. I enquired as to his plans. 'That terrifies me', Elvestad responded. 'It's very easy to start a war. Right now, you have about five or six coastal states – look at how they're struggling and cannot agree', he said, referring to the ongoing fracas over mackerel, herring and blue whiting stocks. His priority was, he told me, to ensure that anything he uses is 'legal and sustainable'. On the high seas, there's always the potential for conflict and criminal activity, he explained. Already, they were having issues with certification. The previous week, in light of the ongoing spat over the mackerel, herring and blue whiting fisheries, MarinTrust – the leading global body that certifies marine ingredients in fishmeal and fish oil – had withdrawn its certification for blue whiting used in fish feed. Another certification body, the Marine Stewardship Council, had chosen to retain their certification for blue whiting, while further assessing the evolving situation.

Catching Vidar a little off guard, I then gingerly broached the issue of the billboard's claims of aquaculture being renewable, curious as to whether there was some rationale I was missing. 'Is anything really renewable today? It's a difficult question', he replied before offering up that 'aquaculture done right – if you have a very low fish-in, fish-out ratio – is actually a net producer of fish. You are making more fish available to the world.' Only, it's never that simple.

Directly in front of the SINTEF SeaLab, on Trondheim's dockside, is the boat terminal, where you can catch a day ferry to Kristiansund, a town further south on Norway's west coast. About two hours into the journey you reach Hitra, a small island in Norway's Trøndelag archipelago. A popular tourist destination in summer, Hitra is also the location of the world's first ever successful salmon farm. It was here, in the summer of 1970, that two brothers, Ove and Sivert Grøntvedt, transformed global food production by placing 20,000 juvenile Atlantic salmon into large floating octagonal cages. Strong and inexpensive, the cages made it easy to feed the salmon while keeping predators at bay. Before long, others began replicating this approach all along Norway's west coast. The following year, in 1971, Norway exported 977 tons of salmon; it now exports over 1.4 million tons, which – according to the Norwegian Seafood Council – supplies the world with 14 million meals each day.

Today, the site at Hitra is owned and operated by Leroy, one of Norway's top salmon producers; on this island, they have a hatchery, where the juvenile salmon live in freshwater for the first nine months, mimicking their natural life cycle, along with a fish farm, where the smolts (young salmon that, in the wild, start to migrate from freshwater to the ocean) are put out to sea in cages and farmed for about fourteen months. At around two years old, the salmon become food, courtesy of a processing plant, also on Hitra. The day after my visit to BioMar, I decided to take the ferry to Hitra to get a closer look inside Norway's fish farming industry and the plans for its expansion. My guides were two locals, Marcus and Brent, who had grown up on the island and spent a good portion of their youths working at local fish farms. They now work for the Hitra Fish Museum, an independent venture funded by Leroy as a public outreach effort in lieu of paying the government an €18 million (US$19.5 million) fish farming licence.

Leroy's Hitra farm is situated beside Big Forest Island, a speck of rock that is neither big nor forested, but that has nevertheless retained its ill-suited name. Going at full tilt of 40 knots in a small RHIB (a rigid-hull inflatable boat) launched from the harbour, we reached the site in just a few minutes and, in spite of a Shipping Forecast that had predicted rain, it was crisp and dry with a light breeze. Directly in front of us were two neat rows of five large floating cages, each twice the size of a football field, and behind them a barge from which several large funnels jutted skyward. We climbed a steep ladder to the top deck. The funnels, Brent told me, were the silos that stored the fish feed; extending out from the barge were ten long white hoses, each supplying a whirling tap at the centre of one of the cages. Dry pellets rattled noisily through the hoses; over a couple of hours they would deliver a constant supply of food – 44 tons in total – to the salmon below us. In each cage were 200,000 salmon; 1.8 million in total. Stretching north and south of Hitra, all along Norway's west coast were 1,000 similar farms, said Brent, holding around a billion fish. All going to plan, by 2050, there will be 5 billion fish in cages lining these shores and filling its waters further and further offshore. For context, Brent reminded me that Norway's wild salmon population is now 500,000 individuals, a number that could squeeze into just two and a half of the cages at Hitra.

Brent's first experience of fish farming had been with his grandfather, who ran a 'farm' off a nearby island. Theirs was a simple set-up – just a net held open with some wooden poles and an anchor to hold the whole structure in place. They'd catch about twenty wild young salmon every year, enough for their own use, and grow them in the net. That was typical until the 1970s, when the Grøntvedt brothers made a success of their larger operation. 'This industry was born out of necessity', said Marcus, who is an historian. 'Fisheries in Norway were really going tits-up in the 70s, and this was the answer. Especially along the Norwegian coast, people were fishermen. This industry saved the coast in many ways – we had to develop because people were moving out of here when I was growing up. It was hard to earn money in these small coastal places.' Brent recounted a similar story. 'In the 90s we had more and more people moving away. Some of the schools closed because there were too few jobs', he said. Now, thanks to salmon farming, the area has been reinvigorated. 'We have a lot of immigrants, especially from Eastern Europe. We have more kids. In the schools, I don't know how many languages we have. We have more activity on the football fields. And the big companies have given a lot of money for building football fields', he told me. 'This is a very important industry, not just for this area, but the whole of coastal Norway.'

Despite the clear benefits of a booming local industry, there are significant problems, one of which is the feed. 'This farm takes up to 40 tonnes of feed per day', said Brent, 'and there are 1,000 of these along the coast and more in other countries', he added, referring to the overseas expansion of Norway's salmon industry, especially into South America, where environmental regulations are more lax. Moving downstairs to a hold inside the barge, I spotted several large bags filled with varied-size pellets. Brent explained that the juveniles in the hatchery are still fed a fish-only formula to help them grow, while the adults have a blend that, these days, is mostly 'terrestrial' or soy-based. 'Now they want to grow this five times. Some of what they are looking for is new resources in the sea – those deep-water fish', said Brent, referring to the mesopelagics of the twilight zone. On a screen in front of us, he pulled up a map from BarentsWatch, an online data tool that showed the fish farms spotting Norway's coastline: they are organised into zones, a method that dictates when farming is allowed and when areas must be left fallow for recovery. They are also colour coded – from yellow through to red – according to their levels of sea lice. These tiny parasites that infest the salmon can ruin the quality of the flesh and, in high enough numbers, even kill their host. According to the colour-coded system, all looked in good order, but I couldn't help but wonder where an extra 4 billion fish would fit along the already-crowded coastline.

Before leaving Hitra, Brent recounted an incident from a fish farm where he'd previously worked. The farm had ordered a delivery of cleaner fish, wild wrasse and lumpfish caught from Norway's coastal waters, which feed on parasitic sea lice. The wrasse and lumpfish are used to clean the farmed salmon of these tiny predators and, once they've done the job and the salmon are farmed, they are typically dispensed with. Owing to a logistical problem, however, the cleaner fish arrived before the salmon smolt were ready to be transferred from the hatchery to the sea cages. With no lice to feed the wrasse and lumpfish, Brent's job was to find protein to supplement their dry feed. 'We got some fishing nets and started fishing. And then we used the fish to bait some crab traps, and then we crushed the crab to feed the cleaner fish', he told me. This kept the wrasse and the lumpfish going until they could feast on the sea lice, which would keep the salmon clean for humans. On hearing this convoluted chain of events – whereby various species of fish and crab were all needed to farm salmon – I could suddenly see why fishing species that in the wild are food for mackerel, but that in cages would feed farmed fish, could seem, to some, perfectly logical – in some twisted twilight-zone sort of way.

THESE DAYS, IT SEEMS superfluous to mention the importance of measuring carbon on land or in the atmosphere, but perhaps it is less obvious when it comes to the deep sea. So consider, for a moment, that the ocean soaks up about a quarter of our fossil-fuel emissions. It holds around fifty times more carbon than the atmosphere itself and around twenty times more than all of the plants and soil on land. By some estimates, the atmospheric concentration of CO_2 would be 50 per cent higher without this carbon-removal service. Owing to the work of people like Ken Buesseler, an ocean biogeochemist at Woods Hole Oceanographic Institution in Falmouth, Massachusetts, the twilight zone is now recognised as the ocean's most reliable pathway for taking CO_2 from the Earth's atmosphere to the deep sea, where it can remain locked up for hundreds or thousands of years. No one yet knows whether, and to what degree, we can fish the twilight zone without wrecking the ocean's carbon storing service. Perhaps more than anyone else, Buesseler can help answer this question. Now in his early sixties, Buesseler has an Irish complexion, ginger hair and an unmistakable New England twang. Over a career spanning thirty-five years, he's been on more than forty ocean expeditions.

On land, Buesseler runs a lab called Café Thorium – a cute name for what is essentially a radiation analysis centre. Buesseler's focus is on the biological carbon pump, a term that describes the variety of ways in which marine animals transport between 2.2 and 6.6 billion tons of carbon each year to the deep sea, some of which happens through the daily mass migration of fish, jellies and plankton that occurs between the twilight zone and the sea surface. In terms of carbon transport, this animal migration between the ocean depths and the surface is a lot like a relay in which zooplankton, small aquatic animals, gobble up microscopic marine algae, known as phytoplankton, acquiring their carbon, and in turn pass it on to carnivorous fish, who eventually excrete faecal pellets that carry the carbon deeper down, or else die and bring it to the deep sea, where it remains locked up.

The larger conduit for carbon, however, is 'marine snow', a rather euphemistic term for ocean detritus. Over land, a snowfall is that rare and cherished experience of untainted nature – an occasional horizontal blur of frozen freshness. Though different in shape and size, each flake is dazzlingly, brilliantly white. In the ocean, snowfall is something quite different. The blizzard that rains down from the sunlit surface is constant; it falls slowly and softly, day and night, over almost every inch of ocean space. From a distance, ocean blizzards have all the scenic wonder of a dark snowy sky over Central Park. But up closer, marine snow is a hodgepodge of pinks, greys and browns, some fluffy and others roughly jagged. Mostly, it's bits of old shell, dead plankton such as diatoms, flakes of fish skin and fish poo. Unappealing as it might seem, this waste from the sea surface is treasure for the twilight zone: without it, there would be no life in this mid-water layer.

The existence of marine snow was first proposed by *Challenger*'s chief scientist Charles Wyville Thomson, who suggested that there was a 'rain of detritus' that fed life further down. In a summary essay on the findings from *Challenger*, written in 1899, one of the expedition's members, Sir John Murray, wrote: 'It has already been stated that plant-life is limited to the shallow waters, but fishes and members of all the invertebrate groups are distributed over the floor of the ocean at all depths. The majority of these deep-sea animals live by eating the mud, clay, or ooze, or by catching the minute particles of organic matter which fall from the surface.' The first person to knowingly see marine snow – and to name it – was William Beebe, an American biologist who in 1930 became the world's first aquanaut. His vehicle, the *Bathysphere*, was built by his co-pilot and collaborator Otis Barton, a wealthy engineer and Harvard graduate with an insatiable appetite for adventure. It had a series of small portholes, a light and a telephone cable to the mother ship.

Together Beebe and Barton made a series of now-famous dives off the coast of Bermuda during which Beebe relayed, to a surface transcriber, his awe at the underwater world in which he found himself. One of his observations was of the constant shower of tiny particles passing by the portholes. An accomplished writer as well as a *National Geographic*–funded explorer, Beebe documented each of their descents with poetic eloquence. During one particular dive, taken on 11 June 1934 at 10 a.m., Beebe described long strings of salps 'lovely as the finest lace', and schools of jellyfish that 'throbbed their energetic way through life'. At 800 feet, he glimpsed a hatchetfish; at 1,000 feet, he noted how the deep sea was like a mirror of the night sky, glowing merrily with bioluminescent creatures, whose existence depended on the constant rain of food from above.

In the decades that followed Beebe's observation, scientists became increasingly interested in marine snow. The Japanese came up with their own word for it – nuta. In the 70s and 80s, scuba divers used dye to mark a starting line in the ocean against which they could measure the descent of the particles and a stop-watch to time them. Eventually, they developed simple traps to capture the falling particles, giving them a rough idea of how heavy the blizzard is at any one point in space and time. They noticed that the deeper the traps, the lighter the snowfall. They eventually were able to show what Wyville Thomson had suspected all along – that marine snow is sustenance for life in the deep, and that hungry animals, mostly zooplankton and fish, were eating the snow as it sank.

It was as a student, in the 1980s, during an expedition to the North Atlantic, that Ken Buesseler devised a simple way of measuring the journey of carbon from the surface to the deep sea, using marine snow. Just like carbon, thorium is an element which exists in different isotopes, or varieties with different chemical weights, some of which are radioactive. Thorium-234, a radioactive element that forms from the decay of uranium-238, is common in trace amounts in minerals found in the environment. If you know how to measure it, you'll find it in harmless quantities on rocks, in soil and in the sea. One of its properties is that it is sticky. 'It will attach to any surface it can find, but the main surface it finds in the ocean is marine snow', said Buesseler, who realised, early on, that by measuring thorium, he could gauge the amount of carbon sinking to the deep sea, in marine snow. He has now replicated this all over the world. From his most recent estimates, it seems that most of the carbon in marine snow makes it to the twilight zone, where it ends up being recycled through the food web – eaten and excreted and so on. Around 220 to 660 million tons of carbon per year (that's somewhere between half and more than all of the UK's annual greenhouse gas emissions) makes it as far down as approximately 1,500 feet, to roughly the middle of the twilight zone. A smaller portion – about 110 to 220 million tons per year – sinks into the abyss, mostly carried within marine snow by the force of gravity. In most parts of the ocean, marine snow that falls below 1,500 feet stays out of the atmosphere for hundreds of years; below approximately 3,000 feet, it stays locked up for thousands.

Working alongside the ocean's creatures to sequester carbon is the 'solubility pump'; this is, essentially, a carbon-carrying service provided by the ocean's physical machinery. It works like this: at the sea surface, atmospheric carbon dioxide exchanges with CO_2 in the ocean to become dissolved inorganic carbon. The colder the water, the more CO_2 is absorbed at the surface. As cold, dense water sinks, it transfers gases such as oxygen, nitrogen and carbon dioxide to the deep sea, and out of the atmosphere. The main benefit of the solubility pump is to get carbon out of the atmosphere and into the surface waters. No matter how carbon is introduced to the ocean, however, the biological processes then take over. Marine snow sinks. Fish, crustaceans and zooplankton partake in daily rituals of visiting the surface to feed and take this carbon with them on their return to the ocean's depths.

In 2018, Buesseler and his colleague, a biologist and innovator named Heidi Sosik, secured $35 million from TED's Audacious Project – the largest single donation ever made to the Woods Hole Oceanographic Institution – to explore the twilight zone. Sosik, who specialises in designing robots and camera systems to investigate underwater life, used the grant to develop new underwater vehicles with better camera systems. Her interest is in the small stuff – fish larvae, copepods and other life forms low down the food web. One of the innovations Sosik uses is the ISIIS (for In-Situ Ichthyoplankton Imaging System), a rapid-fire camera towed off the back of a ship on a sled known as the Stingray and designed to capture imagery of plankton floating freely at sea. The pictures taken by ISIIS are shadowgraphs – ghostly images captured by bouncing light off these creatures, casting shadows over their edges and insides, and revealing details invisible both to the human eye and to conventional cameras. This optical method can capture some of the twilight zone's smallest and most fragile creatures. Many have gelatinous bodies, largely made of water. Capable of withstanding the incredible pressure of the deep sea, at the surface these creatures lose their shape, becoming amorphous and barely recognisable blobs. So, they have to be photographed in situ.

Using ISIIS, Sosik can image organisms living in the twilight zone at 14 frames per second and send the footage straight back to the ship through a fibre-optic cable. The images are delicate black and white depictions of creatures which we rarely see or consider: a pulsating comb jelly, a long colony of salps strung together like a beautifully beaded necklace, and an unruly siphonophore captured mid-dance, its wiry tentacles elongated in pursuit of some unsuspecting prey. Typically, ISIIS images creatures that measure just about an inch across. Also on board the ship is an IFCB (Imaging FlowCytobot), a camera system that Sosik uses to photograph diatoms. These mini-algae are the ocean's biggest source of plant matter, and a common component of marine snow. Encased inside an ornate silicate structure, each diatom is about the width of a human hair. Some look like flowers or leaves, others like shrunken bugs. Displayed like a jumble of rare and unrelated ornaments, it's hard to fathom that they are invisible to the human eye, and part of the intricate machinery that regulates Earth's climate.

TO THE EARLY EXPLORERS, the ocean had unlimited abundance; it seemed there were too many fish and whales in the sea for humans to pose any real threat to their existence. Yet as scientists probed further, and deeper, they realised that not all regions were the same. Some large expanses of ocean contained little life, while others, particularly those along coasts, proved much more productive. The most fertile fishing grounds were often along the west coast of continents, their capacity for sustaining life fuelled by the presence of eastern boundary currents, which bring life-sustaining cold water from the poles towards the equator. As scientists continued to explore, it became clear that there are both fertile and barren ocean plains – as well as everything in between – just as there are on land. The central North Pacific Ocean, for instance, is like a simple grass meadow; compared to a rainforest, there's little nutrition here and so there are fewer plants and animals. It's a quiet environment where growth is slow. By contrast, the Northeast Atlantic Ocean boasts strong currents, which make its surface waters rich in nutrients and conducive to large plankton blooms in spring. It's more like a lush forest. There's not only more food at the ocean surface, but more sinking detritus to feed life further down.

As a particle of marine snow begins its ocean descent, it first passes through the 'euphotic' or sunlit surface waters, a layer where most plants and phytoplankton capture energy from the sun and convert it to food. If a particle of marine snow survives its passage through the sunlit surface without being gobbled, at around 300 feet it enters that nebulous ocean mid-layer, neither light nor dark, known as the twilight zone. Most marine snow that makes it this far stays there, consumed by large carnivorous plankton or small fish that live between 300 and 3,000 feet deep. These plankton and fish in turn become food for whales, turtles, tuna and sharks, and through this chain of consumption, marine snow becomes the lifeblood of some of the most iconic parts of the high seas. Marine snow that begins its journey in the Pacific, halfway between Hawaii and the Baja California peninsula, might, for instance, feed plankton that will feed a squid that will become lunch at a place called the White Shark Café. Hundreds of great white sharks travel from coastal California each December and stay at this shark lair throughout spring to forage and to breed.

Further south, in the eastern tropical Pacific Ocean near the coast of South America, marine snow feeds zooplankton, which is eaten by blue whales. Here, in what is known as the Costa Rica Thermal Dome, a current system caused by strong winds forces deep, cold waters to come up just below the surface. This unique upwelling has the highest concentration anywhere in the ocean of chlorophyll, a green pigment that is present in plants including phytoplankton. At the Costa Rica Dome, the chlorophyll levels are high enough to produce a rich rain of marine snow.

First discovered by American scientists in 1948, it's only recently that the Dome's importance as a marine biodiversity hotspot has come to light. Typically, the Dome sits north of Costa Rica's Pacific coast, in an area called Papagayo, and measures 900 by 300 miles. Its position, shape and size, however, are influenced by winds and by ocean currents. For six months of the year it is blown further out and exists partially on the high seas while portions of it extend towards Costa Rica, Nicaragua, El Salvador, Guatemala and Mexico. It contains an enormous amount of plankton, drawing iconic ocean species at either end of the food web that come here from far and wide to feed and breed. Visitors include tuna, sharks, squid, dolphins and endangered leatherback sea turtles that swim through the Dome after they've nested on Costa Rica's famous beaches. The blue whale, the largest creature in existence and classified as an endangered species, lives here in all its life stages. UNESCO has included the Costa Rica Dome as one of five areas on the high seas – another being the White Shark Café – that ought to be designated as World Heritage sites.

Meanwhile, in the northern Indian Ocean, just southeast of the Arabian Peninsula, marine snow finds its way into a deeper ocean almost entirely devoid of oxygen. The offshore waters of the Arabian Sea, beyond the coasts and territorial seas of Oman, Yemen, India and Pakistan, are warm and salty, and contain the world's largest and most intense Oxygen Minimum Zone (OMZ), a band of water where oxygen levels fall well below the normal range of half to three-quarters of an ounce per 1,000 gallons to below a quarter of an ounce per 1,000 gallons. In the Arabian Sea, the OMZ stretches from around 500-foot depth to 4,000-foot depth. The reason for its limited oxygen is in part due to physical processes, such as currents, that restrict this water from mixing with the surrounding areas. What's more, there is a lot of phytoplankton at the surface here, some of which eventually sinks as marine snow and is eaten by bacteria. In eating marine snow, bacteria consume oxygen, leaving the waters deeper down even more oxygen-deprived. As these waters are warming with climate change, they are becoming increasingly layered, which further restricts mixing, intensifying the Oxygen Minimum Zone.

Far from being lifeless, however, the suboxic waters of the Arabian Sea provide a home for a rich assemblage of unusual creatures that have evolved to survive in such conditions – the giant red mysid, for instance, is an oversized shrimp whose gills have an unusually large surface area, allowing it to absorb oxygen from the water with high efficiency. The deep-sea vampire squid, *Vampyroteuthis infernalis*, has several adaptations that allow it to thrive here, including a low metabolic rate and a preference for munching on marine snow, in stark contrast to the predatory strategies of most other squid. Recent evidence suggests that vampire squid have been using these strategies to inhabit Oxygen Minimum Zones for millions of years. Deeper down in the Arabian Sea are larger creatures that feast on the delights of the OMZ, among them tuna, sharks, whales and ribbonfish – a slim tape-like fish that inhabits the depths but is rarely encountered. Without marine snow, and the species it supports, none of these places would exist.

With so many species feasting, it's easy to see why only a small portion of marine snow, around 10 per cent, gets through the twilight zone. If it does, however, fall to below 3,000 feet, a particle of marine snow might find a final resting place on the seafloor of the continental slope, a jagged incline made of boulders, exposed rock faces, deep canyons and sliding sediments, which stretches from the outer edge of the continental shelf. Or it might first sink further through the water column before making landfall, through the ocean's midnight zone, a lightless realm that begins at approximately 3,000 feet and reaches down to nearly 13,000 feet. This is a place where creatures such as the angler fish, eels and large jellies thrive in the cold dark. By the time marine snow has sunk this deep, the flakes are bigger, having clumped together making their descent more rapid. Even then, the journey can take several weeks. Some of the creatures that live here have adaptations that allow them to catch and eat these larger falling particles. One is the vampire squid, which can swim down this deep, and which uses two long sticky filaments to collect marine snow as it drifts past on the currents. Swarming with bacteria, these clumps of marine snow are also nutritious and important to small and growing deep-sea animals such as eel larvae that feast on them greedily. Beyond the midnight zone, marine snow sinks into the ocean's 'abyss', a deep, dark zone that reaches down nearly 20,000 feet below the surface. Few creatures can withstand the unbearable pressures and near-freezing temperatures of the abyss; those that do thrive here, however, include the dumbo octopus and deep-sea jellies.

As a name, the 'abyss' belies the plurality of life in the dark depths of our oceans, much of which has yet to be explored. The frigid temperatures, lack of light and little food coming from above made many scientists assume, until recently, that the ocean's abyss was effectively lifeless. That view is now being challenged. The density of life in these slow, deep environments is usually low, but it's extremely diverse; around 80 per cent of the animals collected in recent surveys from abyssal plains in the Pacific, for instance, have been new to science. And far from being flat uniform expanses, some regions such as the Porcupine Abyssal Plain in the Atlantic Ocean have gentle undulating hills reaching hundreds of feet high that harbour congregations of larger animals. From what we know, it seems that around 50 per cent of the Earth's surface ends in the abyss, and that just a quarter of the ocean is deeper still. This deeper layer – the hadal zone – reaches over 6,800 miles in the Mariana Trench. It is the domain of scavengers that feed on the falls from above – not just on the few remaining particles of marine snow, but also occasionally on larger wins such as whale carcasses.

On planetary time scales, this chain of consumption is relatively recent. It seems to have begun around 15 million years ago, at a time when Earth transitioned from being a sweltering, high-CO_2 world to one much more conducive to human habitation, with a moderate climate and with less carbon in the atmosphere. Around the same time, the Antarctic ice sheet expanded to its current size, and orangutans and humans diverged from their common ancestor. The gradual cooling of the Earth's surface in that period – by about 7–11°F – has had discernible effects on ocean life. Generally, when the ocean is warm, there's lots of photosynthesis, but there's also lots of bacteria at the surface, which feed on carbon in its solid form, breaking it back down into CO_2 and gassing it back out to the atmosphere. But as the ocean started to cool around 15 million years ago, it became more efficient at bringing carbon, in this case, food in the form of marine snow, to the deep sea. The cold interior of the ocean preserved this food, allowing it to sink for longer and deeper. Over time, this gave rise to new evolutionary deep-water niches, which exploded into the amazing array of unusual life forms that we're just now starting to see, and appreciate.

ONE ARGUMENT IN favour of fishing the twilight zone is that most of its inhabitants, even fish such as pearlsides and lanternfish, are small. As small species tend to reach maturity earlier and reproduce annually, it's theoretically harder to overfish these populations. By comparison, populations of large animals such as tuna, sharks and whales have fewer individuals, reproduce more slowly, and are quickly depleted. For some, such as Norwegian biologist Webjørn Melle, who is based at Norway's Institute of Marine Research in Bergen, fishing for mesopelagics is a safer bet, from a sustainability perspective, than existing high-seas fisheries which target slower-growing fish with longer life cycles, because in theory the stocks should rebound more quickly. Nevertheless, stocks of other short-lived, fast-growing and highly reproductive species, such as squid, have been known to crash without recovery. What's more, the implications of removing these creatures, and their carbon, from the ocean, remain unknown.

Today, seventeen bodies known as regional fisheries management organisations (RFMOS) have responsibility for managing fisheries on the high seas. The oldest of these, the Inter-American Tropical Tuna Commission, was set up in 1949, but many RFMOS are recent. Several were formed in response to the UN Fish Stocks Agreement, a 2001 law intended to better manage transboundary stocks such as blue whiting and mackerel that roam between the high seas and national waters. Even now, however, RFMOS only oversee certain fish stocks or certain areas – tuna in the Pacific, for example, or general fisheries in a defined region such as the Northeast Atlantic. In theory, these bodies have a remit in allocating quotas between member states, reducing bycatch and minimising illicit activity such as poaching. The limitation of this system is that it leaves some of the ocean's richest fishing grounds, such as large parts of the Indian Ocean, with no oversight at all. And as RFMOS only cover commercial fish stocks – around thirty-nine species in total, most of which are tuna and mackerel species – that leaves 95 per cent of all fish species, including mesopelagics, outside the remit of any authority. Essentially, these stocks are free to be harvested, without restriction.

But the problems with fisheries management go further still. Critics say that RFMOS are rendered ineffective by insufficient resources and that catastrophic declines have happened under their watch. The fact is that roughly one-third of the world's fish stocks are overexploited and two-thirds are exploited to their maximum extent, leaving them vulnerable to overexploitation.

Of particular concern are highly migratory species, such as tuna and their close relatives such as billfish, that spend much of their lives roaming the high seas. On average, these stocks have fallen by 60 per cent during the last half century. As a result of overfishing, for instance, Pacific bluefin tuna stocks in 2020 were only around 10 per cent of their historic numbers. In spite of this, in 2021, two regional fisheries management organisations chose to increase the 2022 Pacific bluefin tuna catch, contravening scientific advice. This sort of behaviour is far from anomalous within RFMOS – numerous examples abound.

Overfishing is wiping out many other species too. Longlines, typically used to catch tuna, have a lot of unwanted bycatch, including around 60,000 to 100,000 seabirds each year, as well as turtles, mammals, sharks, rays and 650 species of other fish that get caught up in this gear. Oceanic sharks and rays, for instance, have declined by 71 per cent since 1970, mostly a result of being caught as bycatch. The general collapse in shark and ray populations is probably even more stark than these numbers would suggest, according to one 2019 study, because of incomplete data from some of the worst-hit regions and because fishing fleets were already expanding in the decades before data were being collected. In the Pacific, leatherback turtle populations have declined by 97 per cent over the course of three decades, mostly from being ensnared in longlines, which can stretch for over 75 miles, bristling with hooks. Recent efforts to curb these losses by RFMOS have been deemed ineffective by independent experts. On the positive side, there is evidence that some tuna stocks are recovering from overexploitation; among these are skipjack tuna in the eastern Pacific Ocean, Atlantic yellowfin tuna and albacore tuna, owing to nations introducing stricter quotas. But as one expert, Guillermo Ortuño Crespo, a marine ecologist and fisheries expert with global non-profit, the International Union for Conservation of Nature (IUCN), put it: 'You can cherry-pick success stories, but ... in the larger scheme of things, if you zoom out and look at the high seas ... there's no end to the carnage.'

The scale of this loss is tragic and much of it avoidable. I, for one, assumed that it must have at least had some benefit for humanity: in maintaining economies, for instance, or even shoring up global food security. Oddly, fishing international waters contributes little to either. The vast majority of high-seas fisheries are profitless. Longlining for high-value catch, such as tuna and sharks, still turns a profit, but bottom trawling and squid jigging are, for the most part, economically fruitless endeavours which only survive because they are buoyed up by government support. That the high seas are difficult and expensive to access means that only a handful of nations now bother maintaining a distant-water fleet. For those few, their work is made possible only by offering the industry subsidies – for fuel, gear or for maintaining the fleet, much like the incentives offered twenty years ago in Europe to develop the deep-water fisheries in the northeast Atlantic. Without these subsidies, a staggering 54 per cent of all high-seas fishing trips would be unprofitable.

But if high-seas fishing does little for the economy, what about global food security? Could it be the case that governments are subsidising high-seas fishing to meet the future nutritional demands of a growing global population? As it turns out, no. Offshore fishing supplies just 2.5 per cent of all fish consumed annually worldwide, mostly providing luxury species such as bluefin tuna and Patagonian toothfish, a meaty white-fleshed fish known in the industry as 'white gold', to restaurants in capital cities such as Tokyo, New York and London. With so few benefiting from an industry that takes such a high toll on marine life, some experts have suggested it's time we close the high seas to fishing entirely, and allow inshore fisheries to rebound. The discovery of plentiful resources in the ocean's twilight zone might just make that possibility more remote.

Yet it's only recently that the idea of fishing the twilight zone has been resurrected: the idea would have likely died off if it wasn't for an expedition in 2010, led by a Spanish oceanographer named Carlos Duarte, who is now based at King Abdullah University of Science and Technology in Saudi Arabia. In 2010, Duarte persuaded the Spanish government to fund a global circumnavigation in honour of a forgotten but once influential Spanish explorer named Alejandro Malaspina. Covering 32,000 miles, Duarte's Malaspina 2010 expedition collected reams of data about the open ocean, documented in numerous research papers. The most famous of these findings was a new estimate of the biomass, or fish by weight, living in the ocean's twilight zone. Using an acoustic echosounder – a tool that emits a brief, focused pulse of sound and measures the time taken for it to bounce back as a way of gauging the presence and size of objects underwater – Duarte's research partner, a Spanish scientist called Xabier Irigoien, figured that we'd likely underestimated the amount of fish in the twilight zone by at least a factor of ten. Rather than the roughly 1 billion tons, as previously assumed, there could be over 10 billion tons of fish there. At the upper end, Irigoien estimated it could be as much as 22 billion tons.

This estimate has been hugely influential and hugely controversial. It's also just one estimate that has yet to be confirmed by any other research team, anywhere. At the SINTEF SeaLab, fisheries technologist Eduardo Grimaldo, for one, retains a healthy scepticism. In five years of searching for mesopelagics, Eduardo said they have had just one successful year of fishing. That was in 2019, in the northeast Atlantic. That summer, in June and July, six Norwegian vessels set sail for the high seas and between them caught 1,800 tons of fish. 'The year after, we were operating in the same area, using the same gear, and we caught zero. Well, not zero, but 11 tonnes, which is effectively zero. Same vessel, same gears', he explained, clearly frustrated by the unpredictable behaviour of the fish. Before each trip, he made adjustments to the nets in the hope of catching more, he told me. 'We thought, okay, we are not getting much; maybe we have the wrong nets? And then the company we are working with continued investing money into developing fishing gear, different nets. We went back to those areas, and we never got more. Not commercial catches, at least', he said ruefully. Talking to Eduardo, I got the impression that he wasn't convinced there was a viable fishery at all and if there was, he wasn't sure we should fish it, often reflecting on his experience in Peru. Even if there are more than 10 billion tons, Eduardo said, if you think about it as being spread out across the global ocean, the quantity in each square foot of water becomes infinitesimally small. 'I get angry, I must say, when people still use this number – 10 billion tonnes', he told me. Meanwhile, others remain optimistic, among them Karl Andreas Almås, a special advisor to SINTEF and one of the originators of Norway's salmon expansion plan. He told me that it's a matter of perseverance, of developing the right tools and technology. 'There are hundreds of millions of tons of these fish', said Almås. 'There's no reason to think this number is unreasonable.' Others agree. 'The resource is almost unlimited in many ways', said Melle, the biologist at the Institute of Marine Research in Norway, who is leading a major international effort to verify the biomass and explore the commercial potential of mesopelagics.

Whatever the exact number, if these fish are spread out across the open ocean, occupying a good chunk of every sea on Earth, it will make them extremely hard – and expensive – to catch. Vessels will need to be at sea for a long time, guzzling fuel, using heavy nets and filtering huge volumes of water. 'If this sounds crazy, we're only doing it because we have no other choice. We've overfished inshore stocks', one American fisheries biologist told me. Another possibility is that mesopelagic fish are concentrated in 'hotspots' – areas where there's plenty of marine snow raining down and the conditions are just right. If we fish these hotspots, we'll be competing directly with larger animals, some of which will end up as bycatch. A common problem is large jellyfish that clog the nets. Included here is the Atolla (*Atolla wyvillei*) – a crimson-coloured beast almost 12 feet long considered to be the king of deep-sea jellies. At its favoured depths of approximately 3,000 feet and beyond, a point beyond which red light penetrates, the Atolla can hide in plain sight, evading predators. Every so often, however, the Atolla emits a dazzle of bright blue bioluminescence – perhaps as a warning to assailants or to startle its prey. There's also the possibility, discussed but unverified, of red-listed species such as sharks being caught. 'The trial fisheries have shown there is quite a large amount of bycatch', Dag Standal told me at the SINTEF SeaLab. 'An important aspect here is that the largest amounts of bycatch are from other regulated species, which Norway share with the EU, UK and Iceland, like mackerel, like blue whiting. If we have a large industrial fishery, then the bycatch will turn out to be quite complex and challenging.'

'If this works, it will be an opportunistic fishery', Eduardo Grimaldo points out. As soon as the resource is there, then you'll have to go and get as much as possible before it disappears.'

Whether the twilight zone can be fished sustainably remains unknown, but Norway is persisting. 'I'm very confident that it can be done and we can manage it', said Melle. 'The question of course is, in international waters, what kind of management? That may be a problem, but we should try to have a way forward on it, because if it's not started by Norway, it may be started by some other countries.' Many see that as a possibility, including Ken Buesseler. 'I think if you could make money on it', he said, 'everybody in the world would be doing this.'

3

THE HUNT FOR DARK TARGETS

A FEW YEARS AGO, I travelled to Panama City to meet Bjorn Bergman, a fisheries analyst-cum-ocean detective who was hunting a fleet of suspected pirate fishers on the high seas. It was a mission in which I had become deeply interested. My years at sea as a working scientist had left me keenly aware of the parlous state of fisheries, but later, as a reporter, I sensed that the unclaimed resources of the high seas were attracting criminal opportunists, and that there was a more sinister side to the fishing industry that showed scant regard for human life, let alone marine life.

My first glimpse of the dark underworld of high-seas fishing came in 2018. Back then, I, like most people, thought that modern-day piracy was opportunistic and occasional – more Blackbeard than Gambino in style. But during a visit to the UN City in Copenhagen, at a conference called 'FishCRIME', I realised the scale of these operations. Over a few long days spent in stuffy meeting rooms, I heard from private investigators, government officials, academics and lawyers about the rising occurrence of transnational organised crime on the high seas. 'It's the wild west out there', one delegate told me. 'The ocean is enormous, and people take advantage of that. They can hide in plain sight because they are the needle in a vast haystack.'

There are numerous ways in which fishers take fish illicitly – intentionally under-reporting the size of the catch, targeting waters where no quotas have been set, or poaching directly from protected areas. Some of this is opportunism and some outright criminality. Often called 'pirate fishing', the technical term is illegal, unreported and unregulated or 'IUU' fishing – an awkward phrase that covers the whole gamut of unsustainable fishing, and which, added together, costs the global economy somewhere between $23.5 and $50 billion a year in illicit profits. Given the scale of the abuse and the extremely high incentives for fraud, it shouldn't be surprising that one in every five fish we now eat is caught illegally. All considered, pirate fishing takes a greater toll on national economies and on the planet's health than more well-known instances of wildlife crime, such as the poaching of pangolins, rhino horns and elephant tusks for use in traditional Chinese medicine. But it attracts far less attention. In Copenhagen, Jorge Rios, coordinator of the wildlife crime prevention unit at the UN Office on Drugs and Crime, put it like this: 'In a line-up of a pangolin, a rhino and a toothfish, which one would most people want to save?' It sounded like a bad joke, but he's right. Many marine creatures aren't as relatable as those that we like to visit at the zoo.

The bigger problem, however, as far as the law is concerned, is the violations that go along with pirate fishing. According to INTERPOL, pirate fishing is an example of 'crime convergence' – an inroad to more serious offences ranging from forgery, tax evasion and money laundering, through to slavery, drug running and human trafficking. Typically, fisheries criminals use an outwardly legitimate business as a front for other illicit activities – trafficking drugs while fishing for tuna, for instance – making their misdeeds and their contraband difficult to detect. The big players also usually transit through international waters, often 'going dark', switching off the usual modes of communication that could be used to locate them, even far from land. On the rare occasions these felons are caught, their crimes are seldom punished: as no one owns the high seas, criminal offences committed offshore, such as slavery, come under no one's jurisdiction. In other words, all of the things that make high-seas *fishing* hard to monitor and regulate make a number of crimes equally hard to monitor and regulate.

In the past decade, there have been several high-profile cases, notably that of the Bandit Six, a fleet of six Spanish-owned vessels that spent years illegally poaching toothfish, or 'white gold' from the Southern Ocean. The vessels, which were last known as the *Thunder*, the *Viking*, the *Kunlun*, the *Yongding*, the *Songhua* and the *Perlon*, were frequently renamed – a strategy that the owners used to evade detection – in addition to frequently changing the flag state of the vessels. At one point, four of the six vessels sailed 'under the red and blue flag of Mongolia', an unlikely flag state if ever there was one. The boats were operated by a single Galician mafia-style family, the Vidals, who earned a vast fortune – an estimated $60 million from the *Thunder* alone – by plundering the seas for nearly two decades, using nets that stretched over 40 miles. It took years, and the joint efforts of INTERPOL, national coastguards, and Sea Shepherd – a group of vigilante ocean defenders – to finally catch them. The owners of this rogue fleet had been helped and shielded by an ensemble of seemingly legitimate actors including insurance brokers, banks, agents and flag states. The Vidals laundered money into Spain using a convoluted chain that involved registering their companies in the tax havens of Belize and Panama, selling the fish to a company in Switzerland, having that company sell the fish on to buyers in Hong Kong and Taiwan, who then paid the Vidals' companies in Spain, which invested the funds in all sorts of enterprises, held under shell companies, from renewable energy ventures to real estate. Eventually, the owners were jailed and fined for laundering money in Spain, although their environmental crimes, carried out on the high seas, went unpunished.

Other criminal cases have focused on human rights abuses, with a couple of big investigations, notably by the Associated Press and *New York Times*, putting fisheries crime on the public radar. A 2014 investigation by journalist Martha Mendoza and her team at the AP, for instance, revealed that hundreds of young, poor, uneducated men had been trafficked from Myanmar, Cambodia, Laos and Thailand and were being held, often for years, as slaves in small cages in a tiny hamlet called Benjina, on Indonesia's Aru Islands. There, they were tortured, beaten and, once unshackled, forced to work on vessels fishing the high seas for up to twenty-four hours a day for a company called Pusaka Benjina Resources. The company even had a mass graveyard on Benjina of those who hadn't survived their ordeal. Much of the fish caught and processed by Pusaka Benjina made its way to US consumers through reputable outlets such as Walmart. In 2016, eight people – five Thai fishing boat captains and three Indonesian traffickers – were jailed for their crimes, and the remaining fishermen repatriated. 'Working with organised crime and homicide did not prepare me for what I saw in the fishing industry', private investigator Tim McKinnel told me. McKinnel now works with the victims of fisheries crime who are, he said, 'at the bottom of the food chain.'

Big players, such as the Vidals, usually target international waters, making their crimes even harder to prosecute: as no one owns these waters, criminal offences come under no one's jurisdiction. Typically, these fleet owners recruit as crew young uneducated men from Southeast Asian countries including Indonesia, the Philippines and Thailand. Desperate for work and with few opportunities back home, they are willing to take a chance on fishing the high seas. Many become debt-bonded to their employers, who have recruited them through 'agencies' that demand a fee for a work placement, a trick that keeps them working as slaves under brutal conditions. Unpaid, indebted and fearing for their lives, they keep working. Using forced labour to catch high-value, endangered species, such as tuna, toothfish or sharks, makes money. 'It does make sense; you wouldn't expect boats with crew who are kept as slaves in appalling conditions to respect quotas or the requirement for fishing licences. One enables the other', said Ioannis Chapsos, a maritime law researcher at Coventry University. 'This is what I'd do if I was a criminal', said Jorge Rios of the UN Office on Drugs and Crime. 'It's an obvious crime and it's a lucrative crime, and if no one is willing to put me in jail or take away my money as a fish criminal then why stop?'

LEAVING THE UN CITY in Copenhagen, I understood the scale of the problem. But what could be done? To answer that question, I needed to meet Bjorn Bergman, who works with SkyTruth, an environmental watchdog that uses satellite and remote sensing imagery to hold industry accountable. His position as a data analyst gave him insight into the scale of pirate fishing on the high seas, but also into the associated criminality. Shy and discerning, with dark-rimmed glasses that frame his stubbled face and usually dressed in shorts and T-shirt (a nod to his interest in hiking), Bergman is bookish and sporty in equal measure. By the time I met him in Panama in early 2019, he'd had many successes; he had identified entire fleets of vessels illegally fishing at sea as well as captains enslaving their crews. His work had helped bring these criminals to justice. Now, he was trying to crack an especially tough case, a pernicious and steadily growing armada of Chinese vessels targeting squid off the coast of South America, threatening to topple local fisheries. Bergman had heard anecdotal reports of crew jumping overboard from Chinese vessels near Montevideo, Uruguay, in an apparent bid for freedom; because of these reports, he suspected the fleet was also using forced labour. According to the Environmental Justice Foundation, a UK-based human rights organisation, 85 per cent of Indonesian crew members working on Chinese fishing vessels report, under interview, that they have experienced abusive working and living conditions at sea; the majority have also reported witnessing or experiencing physical violence.

Like me, Bergman had once been a fisheries observer, only – like most observers – he had had a rough time of it. After finishing a degree in biology, he had moved to Alaska to find work. 'It was a job that you could start out in; they needed to recruit people because they had really high attrition. So they were always trying to bring new people into the programme', he told me. The job was varied; at times he'd join family-run operations for trips of three to four days from Alaska's Kodiak Island, which he really enjoyed. He seemed to appreciate the work, and the opportunity. But he endured longer trips, of a month or so on large industrial boats, working some of the ocean's most dangerous fisheries, such as the Bering Sea. 'It was just a really difficult experience', recalled Bergman of one particular trip from Dutch Harbor on Alaska's Aleutian Islands. By most accounts, working as an observer is gruelling. Since 2007, at least nine fisheries observers, including four Americans, have died on the job. Foul play is suspected in a number of these deaths, including that of American biologist Keith Davis who went missing from a tuna vessel, the MV *Victoria 168*, flagged to Panama on the high seas, in September 2015. Not one of the seventeen management organisations tasked with overseeing fisheries on the high seas have regulations to include the safety of their observers at sea; only four of them have a process to follow in the event that an observer disappears or dies.

After several years as an observer, Bergman was done. He thought about becoming a research scientist – by then he had a master's degree in cell biology – but he saw that the non-profit SkyTruth was recruiting analysts to work with fisheries data and decided to apply. Bergman was hired and joined the team in 2014. Founded by a geologist named John Amos, the ingenuity of SkyTruth lies with using satellite data to hold industry to account. The organisation gained acclaim when Amos and his colleague Paul Woods became the first to challenge BP's reports of the extent of oil pouring into the Gulf of Mexico from the *Deepwater Horizon* blow-out in 2010, the largest accidental oil spill in history. One area that SkyTruth had made little progress with was fishing, but they realised there was untapped value in satellite detection of boats with Automated Information System (AIS) beacons. These devices are ordinarily used by skippers to communicate their positions at sea, but their signals can also be used to figure out who is fishing where, and whether they have the right to be there. So, a year after hiring Bergman, Woods and Amos created a spin-out organisation called Global Fishing Watch (GFW) which would detect, track and map the activity of fishing vessels across the ocean. For the first time in history, there was a way of bringing transparency to an industry that had been purposefully opaque.

From his experience on boats, Bergman possessed an uncanny ability to see patterns and anomalies in the AIS data, which – from the start – numbered over 60 million data points per day. He could, for instance, simply look at the ship tracks on AIS and tell from their configuration whether a boat was in transit or fishing – and if it was fishing, whether it was trawling, longlining (a trailing line with baited hooks at regular intervals), or purse seining (creating a net 'cage' around the fish). Using these data, Bergman also learned to identify other suspicious behaviours, such as unusual fishing locations, and to chart a ship's most likely course in a way that most analysts were simply unable to. Amos, who had hired Bergman at SkyTruth, put it like this: 'Bjorn, having been on fishing vessels out in the Bering Sea, was able to look at those data and say: "Hey I know what they're doing right now." But we had to train machines to understand the data, of course, because we couldn't hire enough Bjorns to analyse the 60 million data points that were coming into the system every day.'

In 2016, the data went online, for anyone to use. Overnight, it was possible to track entire fleets moving between fertile fishing grounds. The data aren't perfect, in part because satellite coverage flicks in and out, so the AIS signal can drop, leading to inaccurate data. More concerning by far, however, is the fact that fishers can simply turn off their transponders in a practice known as 'going dark'. Invisible, these dark targets are presumed to be purposefully evading detection. 'At any one time, as much as 20 per cent of the high-seas fleet can be off-the-radar', Bergman told me. It's this dark fleet that they are intent on illuminating. Bergman has managed this with some notable success. In 2016, for instance, he tracked down a rogue fleet of Chinese longliners fishing the southern Indian Ocean with no obvious reason to be there. At the time, Bergman was working for SkyTruth from their head office in Shepherdstown, West Virginia. As on any other working day, he was hunched over his computer, scanning the sea for fishing vessels using their AIS signals and looking for obvious outliers. On the screen before him were 120,000 vessels, all fishing different parts of the ocean. Dense clusters of boats hugged the coastlines of Peru, West Africa and French Polynesia, marking fertile fishing grounds. Further offshore, the boats scattered. It all looked fairly typical, thought Bergman, except for one anomaly – a group of six vessels bunched together in a remote part of the southern Indian Ocean, close to the search area for the missing Malaysia Airlines Flight 370.

He zoomed in for a closer look. He'd been tracking the vessels on AIS. Each AIS signal usually denotes an individual vessel, but Bergman thought that these ones seemed a little too clustered for that. Suspecting that the vessels were up to no good, he raised the alarm. From West Virginia, he blogged about the suspect behaviour. Over in Perth, Australia, Sid Chakravarty – a captain with the direct action group Sea Shepherd – read Bergman's blog. He immediately readied his boat the *Steve Irwin* and, within hours, launched a patrol mission. What Chakravarty discovered, on reaching the southern Indian Ocean, was shocking even to him, a seasoned oceans defender. The entire fleet was using banned drift nets laid out over miles of ocean, ensnaring species such as tuna, sharks, turtles and dolphins. Nicknamed 'walls of death' because of their destructive capacity, drift nets longer than 1.5 miles have been banned by the UN since 1993. The fishers had attached AIS beacons to their long nets so as not to lose them, which explained the pattern that Bergman had witnessed from his landlocked position. 'What we could see was them laying out a long string of beacons and then reeling them back in', said Bergman. The activists intercepted the fleet, hauled in some of their nets and videoed them. After that, the ships scattered and went dark. But a month later, one boat turned on its transponder again, allowing Bergman to relay an updated position to Sea Shepherd. The team ended up chasing the vessel for 5,000 miles across the ocean to the Chinese port of Zhuhai, where the entire fleet was eventually detained and suspended. The fleet owner was fined US$1 million by the Chinese authorities. Having Sea Shepherd there to witness the event had verified Bergman's suspicions. 'We need to corroborate some of what we are seeing, so we can be sure that we're drawing the right conclusions', Bergman told me. Longer term, matching unusual patterns in the data to specific illegal activities could in theory help analysts, such as Bergman, to quickly diagnose future transgressions.

The following year, in 2017, Bergman helped investigate a landmark case, the largest ever haul of illegal shark products in the history of the Galápagos Islands. A UNESCO World Heritage site, the Galápagos is known as the home of some of the Earth's most endangered species, such as the islands' iconic giant tortoises, but it is also a refuge for hammerhead sharks. Bergman received a call in August to say that the Ecuadorian navy had just intercepted a vessel near the islands. It was a Chinese-flagged reefer, or refrigerated cargo ship – with three chilled warehouses on board – fishing international waters, just outside of the archipelago. The *Fu Yuan Yu Leng 999* had repeatedly refused to respond to radio calls from the navy. A helicopter and coastguard boat were dispatched to take a closer look. On board the vessel, the officers discovered thousands of dead sharks and shark fins, including hammerheads and whale sharks.

The captain and crew were arrested, but the authorities also wanted to identify their accomplices. Reefers often operate as transport ships for smaller vessels: on the high seas, they collect the catch from vessels such as longliners, storing them in their refrigerated holds and transporting them ashore, while their associates stay at sea and fish for longer. 'Just having sharks on a boat in Galápagos is illegal, but they also wanted to know how they got them', said Bergman, who set about retracing the movements of the *Fu Yuan Yu Leng 999* prior to its capture. During his trial in Ecuador, the captain named two Taiwanese vessels as the source of the sharks. But Bergman challenged that account. He could see from the data that the reefer had rendezvoused with four Chinese longliners to the west of Galápagos. 'He clearly gave false testimony at the trial', said Bergman. The ship's owner was fined US$5.9 million and the captain sentenced to four years in prison. In 2019, a crew member from one of the Chinese longliners confirmed that Bergman was right; they had indeed offloaded their catch onto the *Fu Yuan Yu Leng 999*. A subsequent study, published in 2021 in the journal *Scientific Reports*, found that the 7,639 sharks on the reefer were made up of twelve species of shark, eleven of which are found within Galápagos waters. This suggests that the vessel was also likely fishing illegally within a marine reserve. Nine of the species on board are classified as vulnerable or higher risk by the International Union for Conservation of Nature, a global body that assesses the conservation status of wildlife. Following the seizure and the legal proceedings, the *Fu Yuan Yu Leng 999* was co-opted into the fleet of the Ecuadorian navy, renamed as *Hualcopo* and, these days, has the remit of guarding the waters around Galápagos.

BUT FOR ALL the cases that Bergman cracked, countless more remained unsolved. It wasn't long after joining SkyTruth that Bergman noticed a Chinese armada laying siege on the

Southeast Pacific Ocean, near the coast of Peru. Every year, the fleet seemed to grow. In 2010, they numbered 104 vessels but by 2014, the fleet had more than doubled; by 2019, there were at times as many as 500 vessels fishing this region, an increase of 400 per cent in under a decade. The flotilla usually 'fished the line', forming a massive fleet along the invisible boundary of Peru's Exclusive Economic Zone (EEZ) waters, which are famed for their continuous supply of nutrients that feed a wide variety of mammals, fish and other creatures such as squid. The quarry of this Chinese fleet is mostly jumbo (or Humboldt) squid, but also Argentine shortfin squid (*Illex argentinus*), which together comprise the world's two largest squid fisheries.

Of the two, jumbo squid (*Dosidicus gigas*) is the main target of industrial fishers. In Peru it goes by the name of Pota, in Chile, it is Jibia, and in Mexico, the 'Red Devil', after its tendency to turn a deep vermilion when hooked. It is a gorgeous animal, possessing an elongated, light pink body flecked with purple and blue. With eight arms and two longer tentacles, it can reach over 10 feet in length and weigh over 100 pounds. Its mantle, or body, is covered in pigmented cells called chromatophores, lending the jumbo squid a remarkable ability to colour-change, shifting from sandy hues to silver to red in just the blink of an eye. It uses camouflage to escape predators underwater, and can convey messages, with vivid flashes of light, to others of its species. The jumbo squid is also a notoriously intelligent and aggressive creature, that hunts in 1,000-strong schools and is known for its cannibalistic tendencies, including a penchant for eating juvenile jumbo squid. For this reason, fishers often use baby squid as bait. The Argentine shortfin squid, by comparison, is a smaller species, with arms shorter than its body, and reaches about 1.5 feet in length. Most often red or orange in colour, it too can camouflage. Both squid species are short-lived with huge reproductive potential: the jumbo squid lives for one to two years and a female can lay a million eggs at a time, and as many as 20 million during her short lifetime. The Argentine shortfin squid lives for just a year, and a female lays about 750,000 eggs. Despite their reproductive success, squid are also highly sensitive creatures whose populations can plummet rapidly when the going gets tough. Variable environmental conditions, and climate change, impact population success from one year to the next, causing massive fluctuations in catch rates. The worry is that heavy industrial fishing, operating without limits, places yet another pressure on squid populations, damaging these species that occupy a vital mid-level role within marine food webs, as both prey and predators.

There are no catch limits for squid in many parts of the high seas, and while the emergence of a squid fishery in the Southeast Pacific Ocean has attracted considerable public attention, it's part of a much wider trend in international waters, where squid fisheries are now booming in a regulatory vacuum. In the past few decades, as finfish populations have crashed, squid fisheries have exploded. Squid populations aren't currently classified as endangered, but conservationists are becoming increasingly concerned by the level at which they are being hunted, especially off the coast of Peru, where 50 per cent of the world's squid landings are caught. It's only in the past thirty years or so that Peruvians have fished these squid commercially. Before then, Peruvians mostly caught finfish, familiar species such as mackerel, hake, and Peruvian anchovy or anchoveta – a tiny silvery-blue anchovy native to the waters off Peru and Chile – as well as the mahi-mahi, also known as the common dolphinfish, a species that sports a long body and a blunt face, with a distinctive domed head. Peruvian anchovy was once the world's largest fishery, with landings off the coasts of Peru and Chile reaching over 10 million tons per year. That was from the late 1960s to 1971, but in 1972, during an El Niño event, the population experienced a near collapse. It rebounded, only to drop sharply again during another El Niño event in 1998. During these climatic phenomena, the ocean of the central and eastern tropical Pacific becomes unseasonably warm, and the surface waters become less productive, all of which leads to a decline in food available for anchovies, and a sharp decline in the stock.

While the anchovy fishery still exists, it now operates at a lower level, alongside an active, and seemingly healthy mahi-mahi fishery. Increasingly, however, locals are fishing squid in competition with foreign industrial fleets. The Chinese distance-water fishing fleet use large vessels with powerful overhead lights to attract squid to the surface at night, and then reel them in with mechanical jiggers, baited with barbed hooks. They first began targeting squid of the coast of Peru in the 1990s; now they are fishing these stocks hard, while also expanding into other regions, such as the Northwest Indian Ocean, the Northwest Pacific Ocean and the Southwest Atlantic. In the four-year period from 2017 to 2020, the fishing effort of the global squid fleet increased 68 per cent, from 149,000 fishing days per year in 2017 to 251,000 in 2020. Over that time, the fleet has expanded its reign in these four areas, using highly mobile vessels that move easily between fishing grounds, large enough to allow fishers to stay at sea for long periods, typically three months to a year, maximising their catch based on seasonal abundance.

The broader context is that in the past six decades, the catch of cephalopods – a term that includes not just squid but cuttlefish and octopus – has risen tenfold, from around 0.55 million tons annually in 1950 to a peak of 5.35 million tons in 2014. The most rapid increases have been in squid fisheries, which now make up a sizeable portion of the landings in some ocean regions, according to the UN Food and Agriculture Organization (FAO). In the Northwest Pacific Ocean, declines in traditional fisheries, such as Japanese pilchard and Alaska pollock in the Northwest, have been offset by a boom in cephalopod landings, most notably of squid, suggesting that squid is becoming a market substitute for traditional fish species. Here, the main targets are Japanese flying squid (*Todarodes pacificus*) and neon flying squid (*Ommastrephes bartramii*).

Meanwhile, in the Southwest Atlantic, Argentine shortfin squid now makes up as much as 40 per cent of the region's annual catch. This overall trend suggests that rather than fishing less, as the statistics on landings of finfish might suggest, we're just substituting fish with species from lower down the food web. According to one analysis, published in 2016, global catches of finfish peaked at 143 million tons in the mid 1990s, and have since been in decline. Conservationists worry that squid are now facing the same fate. Most of the global cephalopod catch is, these days, by large industrial squid jiggers fishing the high seas. China, in particular, now uses its offshore fleet, in part, to pursue squid in other countries' national waters and throughout the ocean with increasing vigour. Compared to the US and EU, which each boast an offshore fishing fleet of around 250–300 vessels, China has a vast high-seas fleet, the exact size of which is unclear. China claims that it numbers around 2,700 vessels, while an independent estimate put the number closer to 17,000 ships, used to target high-seas stocks indiscriminately. As the captain of one Ecuadorian tuna boat put it: 'They just pull up everything!' While their presence on the boundary of Peru's EEZ is perfectly legal, it's seen as aggressive.

The threat to Peru has made this mission personal for Bergman. As a young child, he lived in Peru for a number of years with his family. Later, in high school, he visited for summer vacations with his father who was a *National Geographic*–funded researcher investigating agriculture in the High Andes. 'It's only relatively recently that they started doing this', said Bergman, 'so we have no idea of the implications.' One possibility is that their continued presence will wipe out the squid stocks which not only support larger animals higher up the food web in these waters, but which everyday Peruvians depend upon for their food and work. Peru and Ecuador each have large fishing fleets and are highly dependent on seafood, both for sustenance and for export. In 2018, the two countries captured 5 million tons of fish, a catch on a par with that of the US, but even so only about a quarter of what China harvested from the sea. The sheer size of China's distant-water fleet poses a genuine threat to locals but more concerning still is the fact that the vessels often go dark, a sign that they might be also taking in illegal catch or fishing protected waters. There are other offenders, for sure. China, however, is now the single nation most involved in overfishing, illegally intruding into protected waters, targeting endangered species, falsifying catch records, and facilitating crimes, including human slavery, at sea.

AT SKYTRUTH, BERGMAN was keen to figure out if the Chinese fleet fishing the line on Peru's EEZ was operating not only aggressively but illegally – by fishing inside protected areas, for instance, or using forced labour. This would prove far more challenging than his previous missions, such as identifying an individual transgressor like *Fu Yuan Yu Leng* 999 or even a small fleet such as those in the western Indian Ocean. This was a much bigger beast, an armada of illegal operators who had figured out how to spoof the system by going off the radar. With no authorities or rules to limit their catch, the 6,000 or so crew operating the Chinese fleet of squid jiggers could stay in the South Pacific for a year at a time, sustained by large cargo reefers that would routinely bring them fuel and offload their catch. While spending prolonged periods at sea is perfectly legal, 'transhipment' or movement of a boat's catch between vessels, is among the most common indicators of crime in international waters. For a start, it's an easy way of hiding poached fish, and other contraband, among legal catch. And because it allows for ships to spend long periods away from land, it is closely linked to slavery at sea – that's to say that it allows captains to keep their crew by force at sea for long periods. As a practice, transhipment is closely linked to pirate fishing and to human rights abuses on the high seas.

To stop these criminals, Bergman would have to change the rules of the game – and for that he needed Panama's help. That was how, in March 2019, I found myself en route to one of the world's most famous tax havens. I was there to witness Bergman and his colleagues from Global Fishing Watch persuade Panama to hand over data that would help them catch fishers flying the Panamanian flag, a move that could expose some big offenders, including the Chinese fleet of squid jiggers. The rationale for approaching Panama was this: to identify a vessel at sea, Bergman either needed to see it with his own eyes, have reliable AIS data or, the next best thing, VMS (vessel monitoring system) data. VMS is a system comparable to AIS, used for routinely logging a ship's position, name and call sign. Most nations mandate their larger domestic vessels carry VMS on board, but the system is hugely expensive and the data proprietary to governments. If Bergman had a vessel's VMS data – which he hoped to acquire from Panama – he'd be able to identify and track it, even if it went dark on AIS. Bergman suspected that some of the larger supply vessels – the reefers – in the South Pacific were flagged to Panama. With the VMS data, their locations would be known, bringing Bergman one step closer to unravelling this tangled web of criminality.

At the time of my visit, Panama's government was still reeling from the humiliating exposé that has been dubbed the 'Panama Papers'. Long known as a place where the very-wealthy can evade and avoid taxes and sanctions and hide their assets, even those in the know were blind to the scale of fraud being facilitated, in some cases, by this tiny isthmus. Then in April 2016, the German newspaper *Süddeutsche Zeitung* blew the whole scheme wide open with details of almost 12 million documents leaked from a Panama-owned law firm, Mossack Fonseca. The documents revealed how 'MossFon', over a period spanning almost forty years, helped celebrities, kings, presidents and Mafia mobsters, among others, hide their money in anonymous shell companies, headed by sham directors – all from a relatively inconspicuous office in Panama City. The Panama Papers revealed that this single law firm had helped to stash almost $23 trillion, offshore, in secret accounts.

Far from anomalous, MossFon and its misdemeanours were simply the pinnacle of a pervasive culture of corruption that was actively encouraged in Panama and that, over the years, had developed many different guises. One of those – perhaps less famed than outright money laundering – is Panama's 'open register'. More commonly called a flag of convenience, Panama's open register means that any ship can fly its state flag, regardless of where it was built, or the country in which its owner – any indeed any crew member – lives. The whole process can be completed online in just eight hours; you needn't even visit Panama to register your vessel there. The advantages of this include low or no income tax, the ability to hire cheap labour, flout environmental regulations and – through shell companies – conceal the true ownership of a vessel or even an entire fleet. Panama isn't the only nation with a flag of convenience; indeed, some open registers such as Mongolia are landlocked. Of the 100,000 seafaring vessels currently in existence, 73 per cent are flagged to an open register, of which Panama is the world's largest, with over 8,000 vessels. Of those 8,000 ships, just 275 are involved in fishing, but crucially this includes the international fleet flagged to Panama. In the wake of the MossFon scandal, Panama was understandably keen to reclaim some air of respectability.

Over several days in Panama City, I had the rare opportunity to peek under the hood of an organisation that is, bit by bit, dismantling one of the largest and most complex criminal networks on Earth – not by staking them out, but by analysing data collected using state-of-the-art technologies. In trying to expose the behaviour of the Southeast Pacific fleet, Bergman wasn't alone; Global Fishing Watch now numbers twenty-two people, dispersed worldwide. Bergman had been using AIS to decipher the fleet's movements, but because the vessels were spoofing the system – either by going dark or relaying false positions – the team needed another way of tracking them. One was a relatively new technology that made use of a National Oceanic and Atmospheric Administration satellite sensor called the Visible Infrared Imaging Radiometer Suite (VIIRS) and could detect sources of light from space at night. Paul Woods had used it at SkyTruth and realised it could pick out the squid jiggers, because they use bright yellow, green and blue lights to attract squid to the surface at night. By 2018, Global Fishing Watch had VIIRS data for the whole ocean, allowing Bergman and the rest of the team to see many high-seas squid jiggers that had, up to that point, been entirely dark. But in a fleet 300-strong, like the one fishing the South Pacific, not all boats are squid jiggers. Some are reefers, whose job it is to restock the fishing boats and to tranship the catch. To identify the reefers, they needed another data source, synthetic aperture radar, or SAR, imagery. The advantage of radar imagery is that it can pick out large objects with metal hulls, such as reefers, that are neither broadcasting their position nor displaying visible lights from space. Beyond these, there's also optical imagery, which is a term for forensic-level satellite photography, such as from DigitalGlobe's *WorldView-3* satellite. These images are expensive, but they are the only option if you want to capture an image of a known target that could be used as evidence in a criminal trial – a photo of a fisher such as the *Fu Yuan Yu Leng 999* hauling in a shark, for instance. 'You can do some work by association', explained David Kroodsma, head of research at Global Fishing Watch. 'You see a fleet of several hundred boats going somewhere on radar, then you look at the AIS and you only see thirty or fifty, right? And so you can often infer from those what the other ones are doing. You have to be smart about it, but you can kind of piece it together from different sources of information – AIS or VMS, VIIRS, SAR and optical. Using those, you can put together the whole story about what's happening in a region.' Proving foul play, however, is another matter entirely.

It was a hot March morning in Panama City when Global Fishing Watch presented their case for needing Panama's help, and its VMS data. Officials from several other nations – Costa Rica, Colombia and Ecuador – sat in on the proceedings. By that afternoon, Panama had signed a memorandum of understanding agreeing to share historic and future data. Over the days I spent in Panama, I grew to admire Bergman's perseverance and his flexibility – this case had stretched him far beyond his role as a data analyst. At times he had to play the diplomat, addressing secrecy jurisdictions on the need for greater transparency. It wasn't the only unusual move he'd made. About six months previously, he had tried confronting the Chinese fleet directly, hitching a ride from Panama City with Sea Shepherd on the *Brigitte Bardot*, a vessel named after the actor and long-time Sea Shepherd supporter.

Using all of the tools at his disposal, Bergman had pinpointed the exact location of the fleet, to an area about 700 miles west of the Galápagos Islands. After a week at sea, he came face to face with the cabal he'd been long pursuing. Packed cheek by jowl were 300 vessels, all densely clustered into one of the ocean's most lucrative fishing grounds. The flotilla shone brightly like some great heaving metropolis, each boat glowing iridescent green and yellow from the lights used to attract their quarry. But while Bergman sensed foul play, even in confronting the fleet he had no firm evidence of any wrongdoing. And so disappointingly, his trip yielded nothing – not one obvious infraction that might cause an activist to board a vessel nor indeed any evidence that could be useful in an investigation. Now, however, he had Panama on side. Not only would their cooperation potentially shine a light on Bergman's dark targets, but it would also make him more confident about the identity of vessels located on AIS. He seemed relieved and elated.

Within a few months of the Panama deal, Bergman and the team at Global Fishing Watch started to help the authorities to identify wrongdoers flying the Panamanian flag. The first of these was an 825-ton rust bucket called the MV *NIKA* with a long track record of illegal fishing. Since 2006, its owners had changed both its name and flag state seven times, in a bid to cover their tracks. More recently, it had been kicked out of Russian waters for poaching crab. By June 2019, it was on INTERPOL's wanted list of outlaw vessels. One of Bergman's colleagues – a data analyst named Imam Prakoso – had been closely tracking the MV *NIKA* for several months before its eventual capture. In July 2019, it was making its way through the Strait of Malacca when it was apprehended by the Indonesian coastguard. The crew tried their best to evade capture, falsely reporting MV *NIKA* as a cargo vessel and then, finally, going dark. But as the vessel was registered to Panama, Global Fishing Watch had another option – they could retrace its movements on VMS. As the vessel moved towards Indonesia, they relayed a series of instructions to INTERPOL and to the local authorities, allowing naval patrols to identify rendezvous points where they could intercept the illicit vessel. Finally, at 8:06 a.m. on 12 July, MV *NIKA* was in the net.

Following Global Fishing Watch's success with Panama, other countries, such as Chile, followed suit, making their VMS publicly accessible. Before long the team hope to have VMS data for the whole ocean, and within a decade, it's possible that rogue fishers will have nowhere to hide. China has also bowed to international pressure. In 2020, China introduced a new fisheries code for the first time in seventeen years, including new laws to be applied to its distant-water fishing fleet. These rules include a blacklist system for boat captains and company owners of vessels found pirate fishing, a commitment to include observers on transhipment vessels and a legal requirement for ships to broadcast their position every hour. In some cases, such as the *Fu Yuan Yu Leng 999* – the vessel found off the Galápagos Islands with the largest ever haul of sharks – boat captains can now be banned entirely from the industry. Of particular relevance to the Chinese squid jiggers, however, is the introduction of China's first ever voluntary seasonal closures on the high seas. Announced in June 2020, the closures cover the main spawning grounds of the jumbo and Argentine shortfin squid, a move that should, in theory, allow their stocks time to reproduce and recover. In 2022, China extended these measures to cover the Southeast Atlantic and the Northwest Indian Oceans. Intended as a sign of China's willingness to rein in its oversized offshore armada, the new rules were greeted internationally with a mixture of fanfare and cynicism, with some critics arguing that the closures happen largely outside of the squid fishing season.

Bergman's mission isn't over though. As I left Panama, I reflected on the dedication required to bring an end to pirate fishing; identifying one rogue vessel such as the *Fu Yuan Yu Leng 999* seemed easy compared with figuring out the transgressions of an entire fleet that was hiding dark vessels. Bringing the Vidals to justice had taken more than a decade and the longest sea-chase in maritime history. Even then, many of the crimes they committed on the high seas went unpunished. How many more Vidals or *Fu Yuan Yu Leng 999*s still sailed on unchecked? I wondered at the chances of ending the plundering of the Southeast Pacific by an entire maritime armada. For the time being, these rogue fishers are still at large, even if the net is closing.

4

TREASURES FROM THE DEEP

NOT LONG AGO, I purchased a rock from an online retailer named Jensan Scientifics LLC, in Iowa. Five days later, the rock arrived boxed and heavily swaddled in tissue, with a typed note describing the typical size and composition and the location from where the sample had been taken. In appearance, it was rather bland with a dull exterior. It had none of the glassy, lustrous sheen of obsidian or the rich texture of a veined marble. Despite its simplicity, this was no ordinary rock; it was a 'manganese nodule' which, the retailer assured me, had been collected in a remote part of the Pacific Ocean called the Clarion-Clipperton Zone. It was unclear when or how this abyssal excavation at around 15,000 feet had happened, or what the proceeds of my purchase would fund. Assuming it was all in the name of science, I parted with $147, plus another $30 for the shipping. The note, stuck fast to the lid of the box, informed me that manganese nodules also contain other, even more valuable metals including cobalt, copper, nickel and platinum, and that the nodules often reach 4 inches in diameter. Unwrapped, my tidy specimen measured just half that. Turning the rock over in my palm, I caressed its black, bubbled surface.

In my hand, the rock was lifeless and cold. In its natural environment of the abyss this nodule would be one of trillions scattered over the soft seafloor, offering an occasional hard surface for creatures – from carbon-sequestering microbes to stalked sponges – to cling to. Over time, these attract other animals; octopuses, for example, lay eggs in the sponges that stick to the nodules. Each nodule starts out as a discarded object – perhaps a shark's tooth or the remnant of a seashell. Over a long time, metals gather and form a crust around this node, allowing it to grow slowly – just 0.5 to 4 inches or so every million years. My mail-order sample, which I received in just five days, would have taken somewhere between 500,000 and 5 million years to form.

If any pursuit has come to signify the steady march of industry into the ocean's unclaimed territory, it is seabed mining. In many ways, it's a familiar story. Mining – one of the world's most notoriously destructive practices – is now likely to descend on one of the ocean's most fragile ecosystems. The closest comparison we have that allows us to imagine the damage deep-sea mining could wreak is bottom trawling, which has left visible, physical scars along the ocean's continental shelves. But mining has the potential to alter the deep sea on an unrivalled scale: by removing vast quantities of rock, leaving behind toxic waste and permanently altering ecosystems, much as it has on land. Aside from the physical scars, there's the risk of noise and light pollution, which can interfere with species' ability to navigate and communicate. There's a risk of local extinctions, in some cases of animals that have not yet been discovered. Mining will stir up great clouds of swirling sediment that could spread hundreds of feet to several miles beyond each mining site, burying and suffocating creatures in a sort of slow-motion volcano. Waste from mining, discarded back into the ocean, could cause havoc for free-swimming animals, among them commercial fish species, and introduce toxic chemicals to our food supply. There's also the possibility that it will interfere with the ocean's ability to store carbon. Yet proponents claim that seabed mining is a rare opportunity to harvest an untapped supply of the precious ores we need to power the electric cars, solar panels and wind turbines of the future. They also contend that it could be the start of ethical mineral extraction, profiting developing nations, doing away with child labour and leaving a better environmental legacy than on land.

For the time being, no one has commercially mined the deep sea, anywhere. There have been numerous attempts by companies and nations, but the nascent industry has been beset by problems including high up-front costs, the historical low value of deep-sea ores and a lack of rules, which has – in turn – made investors wary. Despite all of this, the industry regulator – the International Seabed Authority (ISA), established by the UN – is now pushing hard to finalise a seabed mining code – essentially an industry rulebook – and if industry has its way, mining could begin in late 2024. Meanwhile, ISA has awarded thirty contracts to allow prospective miners to explore the mineral wealth of an area totalling around 400,000 square miles. These contracts cover three types of deep-sea mineral deposit: cobalt-rich crusts, seafloor massive sulphides and manganese nodules. The first of these, cobalt-rich crusts, line the flanks and summits of submerged extinct volcanoes – called seamounts – which rise thousands of feet above the seafloor. The crusts accumulate over millions of years and are packed with high-value metals, such as cobalt, platinum and molybdenum. The second type, seafloor massive sulphides (SMS), are found at hydrothermal vent sites. These deposits are formed when hot springs (up to around 650°F) rise through the seafloor and precipitate leached metals, including zinc, nickel, copper, gold and other rare elements. In recent years, industry has shown that it's feasible to mine sulphide deposits, but overall there's been less interest in cobalt crusts, which are technically more challenging to access. Most contracts are to explore the potential for harvesting manganese or 'polymetallic' nodules from the Clarion-Clipperton Zone (CCZ), a vast abyssal plain that stretches from Mexico to Hawaii and reaches depths of nearly 20,000 feet. Faced with the imminent arrival of this nascent industry into the ocean's abyss, there are plenty of fears of the damage it might cause, but few studies, leaving scientists scrambling to collect data.

The sample sent from Iowa wasn't the first manganese nodule I'd seen. Several months earlier, I'd spied one briefly on a visit to London's Natural History Museum in South Kensington. It was a winter's day in early 2019, and I had arranged to visit Diva Amon, a Trinidadian biologist who, at the time, was working at the museum's Darwin Centre. Built in 2009, the Darwin Centre is a colossal white cocoon encased in a glass atrium. The unconventional structure now houses the museum's 17 million entomological specimens and 3 million botanical specimens, all carefully catalogued in cabinets that, side by side, would stretch for over 2 miles. The Darwin Centre also acts as the museum's research facility, providing lab space for 220 research scientists, of whom Diva Amon was one. Her work focuses on deep-sea biology, especially on 'megafauna', a term that evokes images of large beasts such as coelacanths and whale sharks. In the deep abyss, however, this includes any creature that can be identified in a photograph – it could be a starfish, a brittle star, a sea cucumber, a jellyfish, or even the gummy squirrel (a species of sea cucumber that looks something like a cross between a shoe and a banana).

Amon, who in 2017 founded SpeSeas – a non-profit dedicated to raising ocean awareness – has since become something of a celebrity in the deep-sea science world, starring, for instance, alongside Will Smith in the actor's 2021 documentary series *Welcome to Earth*. In 2020, she was named a National Geographic Emerging Explorer, a title bestowed upon outstanding young researchers whose work is likely to change the world. Amon's first research position, at the University of Hawaii, took her to the CCZ, where she was surveying a site earmarked for mining by UK Seabed Resources, then a subsidiary of the arms giant Lockheed Martin. From there, she moved to London's Natural History Museum where, at the time of my visit, she was working in a lab headed by Adrian Glover. A British biologist, Glover is an expert in sea worms known as polychaetes, a group that includes the lugworm (mostly known for the coiled castings it leaves on British beaches) and the sea mouse (a furry worm not dissimilar in appearance to a flattened hedgehog). Now at the University of California, Santa Barbara, Amon's current research is focused on understanding the deep sea, including the CCZ, but not on collecting data for contractors.

Even now, much of the deep sea remains a mystery to us. That's in part because the abyss is a world of extremes. Depths reach several thousand feet, temperatures near the seabed in many places hover near 32°F, there is next to no light, and pressures can exceed 14,500 pounds per square inch, equivalent to having a couple of elephants standing on your big toe. Despite the fact that scientists have been venturing to the CCZ for the past fifty years, surprisingly little is known about what lives there. It's a huge area, so big that if you laid it out across North America, it would stretch from California to New York. Research such as that carried out by Amon and Glover has been given the green light because of industry's interest in exploiting the region. What they're finding is that much of what lives here is both rare and limited in range: by one estimate, 90 per cent of the species collected on surveys of the CCZ have never been seen before, by anyone on Earth. Scientists suspect there are thousands of species that live only in this region, Amon told me. This means that it would be easy to cause the global extinction of a species, or even several, without so much as noticing it's happening.

I peered through a microscope at one of Glover's polychaetes, another specimen that had yet to be described, identified and added to the museum's ever-growing catalogue. Before I left, Glover asked if I would like to see a manganese nodule. Brimming with excitement, I surveyed the dull rock, as barren as a lump of coal – my first glimpse of concealed riches from another world.

THE FIRST SCIENTISTS to uncover evidence of ore-rich rocks in the deep abyss were those on HMS *Challenger*. The scientists noticed them in deep waters southwest of the Canary Islands when heaving up a quantity of dead coral coated in black specks. Around 2,000 miles to the west, they again hauled up a collection of strange fragments – these ones metallic and scattered among larger black nodules, some the size and shape of walnuts and others as big as a grapefruit. The scientists found that the nodules, once sliced in half, all had the same characteristic concentric circles inside, like trees, or as though layers had been laid down, one on top of the other.

Writing in the prestigious journal *Nature*, the chief scientist on board HMS *Challenger*, Charles Wyville Thomson, queried whether they might have accreted these layers over millennia in much the same way that pearls build up in layers around a speck of grit within an oyster shell. These sightings were just the first of many by the *Challenger* crew of what are now called manganese or polymetallic nodules, and recognised as a potential source of valuable rare earth metals. At the time of the expedition, the nodules were considered an oddity from a strange unknown world, and were sent to museums for people to marvel at, much as lunar rocks were a century later.

It wasn't until the 1960s, however, that any thought was given to their potential value and to who exactly owned these precious minerals. The question was this: was the seabed, and its valuable mineral resources, a no-man's-land or common property? According to sixth-century Roman law, common spaces – including the air, the ocean and the seashore – could be governed in two main ways. *Res nullius* were spaces that were owned by no one, but that could be claimed if someone discovered and occupied them. Philosophically, *res nullius* is the same as the free seas. The other possibility, according to the Ancient Romans, was to declare common spaces as *res communis*, or common property. These areas – such as national parks – can't be appropriated, because they are already owned by everyone. Given that the open ocean was nobody's, it seemed safe to assume that the same rule applied to the seafloor below. But in the 1960s one man begged to differ. His name was Arvid Pardo. As representative to the UN for the tiny island nation of Malta, Pardo was the appointed spokesperson for a population of roughly 300,000 people, who inhabited an area of land measuring just over 115 square miles. He was, by all accounts, an unlikely figure to chart a new course for global governance. Yet that's exactly what he did.

In 1967, the little-known diplomat made an emotional three-hour speech to UN delegates in New York, pleading with them to consider the seabed as the 'common heritage of mankind'. The idea, which took its lead from *res communis*, decreed that certain spaces should be commonly owned and that their resources should benefit everyone on Earth. Pardo worried that international waters were being ruled as free seas, and that there was nothing to stop rich nations plundering the deep seabed, once they developed the technology. He wanted to stop a subsea land grab – most likely by the US and USSR – while also addressing what he saw as growing inequality on Earth.

A number of factors worked in Pardo's favour. For one, he became a diplomat at the time when there was a growing belief in technology, epitomised by the space race, but also escalating concern about the expanding gap in income between rich and poor nations. It was Pardo's genius to tap into these fears and offer a solution. With the US and the Soviet Union locked in a competition to land the first astronaut on the moon, the deep sea was shaping up as a parallel kind of frontier, a vision portrayed on TV shows such as the sci-fi drama *Voyage to the Bottom of the Sea* and Jacques Cousteau's adventures.

Not only was the Cold War underway, but developing countries worried that a battle for potentially lucrative ocean resources might lead to full-on US–Soviet conflict. Pardo managed to tap into these aspirations and anxieties, and offer a solution. There were other factors that weighed in his favour. Malta was politically neutral territory, making Pardo's message credible. And his motives seemed clear: by that stage, Pardo had a reputation as an advocate of peace and equality, having experienced harsh injustices in his own life, including the untimely death of his parents and brother and a period in prison for protesting fascism. In making his heartfelt plea to the UN, Pardo had a grand vision, a sort of new ocean order that would right the world's wrongs. He imagined a global body that would oversee not just deep-sea mining, but oil exploration, scientific research, military operations and environmental protection across the high seas, right up to the edge of each nation's territorial waters. Pardo saw the seabed as commonage, but he went further and proposed that any profits derived from its resources be shared among nations.

In truth, Pardo knew precious little about the ocean. In fact, it was at a cocktail party in New York, during his tenure as ambassador to the UN, that he first became aware that states were interested in mining the seabed. But his lack of expertise didn't stop him from airing his utopic vision of the 'the womb of life', as he called the deep sea. He imagined a future underwater world in which humans would live, protected by air bubble curtains at depths of around 1,500 feet, where they would farm fish and grow all food, apart from luxuries like fruit. Dolphins would act as sheepdogs, herding fish stocks, and concentrated fish extract would meet the growing global demand for protein, at a cost of less than 1 cent per child per day. This subsea nirvana would be financed by mineral extraction, an idea that made sense to Pardo, who took his lead from an American mining engineer, John L. Mero, in assuming that mineral-rich ocean rocks were easy to harvest and essentially renewable, growing faster than they could be harvested. Despite the fierce opposition to Pardo's plea, the UN took note and, in 1970, declared that the international seabed should be treated as common heritage, the benefits of which must be shared. The precise details of how this would work, exactly, were to be decided at a later date, but the principle was established.

By the time Pardo won his plea with the UN, however, the prospects for deep-sea mining had started to wane. Funds for exploration dried up, in part because the UN had started to question who, exactly, owned minerals extracted from the seabed in international waters. What's more, in stark contrast to Mero's prediction, extracting these ores proved a huge feat. Exploratory research showed that it was technically possible to harvest minerals from the abyss, but only on a small scale. Designing equipment that could withstand deep-sea pressure, for decades at a time, was an entirely different prospect. With little demand for minerals that could be easily sourced on land, no one was willing to part with the huge sums required, and the business case for deep-sea mining was suddenly quashed.

It took thirty years for commercial interest in seabed mining to re-emerge, and when it did, well into the 90s, it was for a reason that seems ironic: tackling climate change. Companies with an interest in mining the international seabed now claim that we'll soon run out of rare metals – such as manganese, cobalt, copper and nickel – currently needed for building green technologies such as solar panels and electric car batteries, unless we mine the deep sea. By some estimates, land reserves of cobalt, a metal essential for extending the lifespan of rechargeable batteries, amount to about 7.7 million tons, comparable to the amount estimated in the Clarion-Clipperton Zone. The story is similar for nickel. There are about 84 million tons of nickel left in reserves on land, roughly the same amount as in the CCZ. The same again for copper: billions of tons may exist in the deep sea, enough to ensure supply for centuries to come; even easily accessible copper deposits are enough to match annual production from land-based mines.

According to the World Bank Group, we'll need over 3 billion tons of minerals and metals to deploy the wind, solar and geothermal power needed to avoid 2°C (3.6°F) of warming, globally recognised as a level of climate change that is dangerous to humanity. For minerals such as lithium and cobalt, this could mean an increase in demand of 500 per cent by 2050. But not everyone is convinced of an impending metals shortage, nor that, in the event of one, deep-sea mining is the obvious solution. Indeed, the Institute for Sustainable Futures, in a 2016 report, concluded that a global transition towards 100 per cent renewable energy could take place without deep-sea mining, as the projected demand in metals could be met with land-based reserves. It's also worth remembering that as promising as deep-sea reserves sound, the fact is that these are guesstimates: gauging the total amounts of metal in the deep sea, and especially those that are easily recoverable, is extremely difficult.

The other contention from industry is that deep-sea mining is an ethical venture that will benefit humanity on the whole. Since the UN declared the international seabed as common heritage and later, in 1994, created the International Seabed Authority (ISA) as the body tasked with regulating deep-sea mining, the mineral resources of the high seas have legally become shared property. If ocean mining happens in international waters, the profits should be shared among all, including developing nations. In truth, seabed mining is still a money-making venture, but it's now been given an ethical and green veneer that are being pitted, by industry, against emerging environmental concerns.

MY VISIT TO London's Natural History Museum taught me a lot about life in the deep abyss but nothing about the impact of mining. Wanting to know more, a few months later, I found myself heading to sea on a Spanish research vessel named *Sarmiento de Gamboa*, where my host was Henko de Stigter, a Dutch ocean scientist at NIOZ (pronounced knee-oz; in English NIOZ is the Royal Netherlands Institute for Sea Research). By the time we met in person, Henko and I had spent months trading emails discussing the trip, and the possibility of my participation. I had spoken to many scientists and conservationists who were aghast at the prospect of mining the seabed, but here I would join a group of industry engineers who had designed a prototype deep-sea mining machine, a beast named Apollo II, that at full scale could suck precious ores from the Pacific abyss. Henko's team of scientists would document the likely harm this emerging technology could cause to life in one of the most remote parts of our planet. Over the few days I spent with Henko, he expressed excitement about being involved in a new frontier of knowledge. 'I'm a geologist, so I'm intrigued at what is down in the Earth', he told me. 'There's part of me that knows there are treasures in the deep – not the minerals but the discoveries to be made', he said.

Others on the team expressed hope that deep-sea mining could offer a better way of sourcing the raw materials we need for industry. 'When you look at mining now – it removes forests and villages', said Ralf Langeler, then the chief engineer on Apollo II. 'By comparison, deep-sea mining doesn't look so bad.' Some commented on the unknown harm it might do. One of Apollo II's creators told me: 'If we mine on land the damage is contained, but in the sea we simply don't know yet.' Regardless of their views on the scale of impact, all seemed confident that deep-sea mining would happen soon. It was simply a matter of time.

I arrived on a Sunday in August to the Port of Málaga, from where I was collected and brought to the huge Spanish-operated research vessel. As it was a Sunday, and perhaps because of my arrival, they'd arranged a barbeque with music and beer on the ship's main deck. After being shown to my berth, I was escorted to the main deck to join in the festivities, and introduced to the team – Henko's scientists from NIOZ, the tech crew who operated the underwater camera and video systems, and a group of researchers from the Spanish National Research Council in Madrid called CSIC (pronounced, somewhat disconcertingly, 'sea-sick'). CSIC owned and operated *Sarmiento de Gamboa*, which would be my home for the next few days.

Henko is tall (in a way that only the Dutch seem to achieve), thin and sports a mop of wiry greying curls. An ocean geologist, one of Henko's interests is the impact of mining the deep sea, but his love of rocks began long ago, during a period living in the Congo while his father taught at a local university. He spent the mornings being home-schooled and the afternoons amethyst-hunting in the driveway with his sister. 'There were gems everywhere, covering the ground', Henko recalled, 'but the people, they were cripplingly poor.' Today, the Congo is one of the world's mining hotspots – not only for precious gems such as amethyst but for the rare earth metals that seabed miners hope to extract from the Pacific abyss. Seventy per cent of the world's cobalt, for instance, is mined in the Congo, mostly by large multinationals, often using cheap child labour in unsafe conditions.

For Henko, this was one reason to consider mining the deep sea. For this project he had teamed up with Royal IHC, a Dutch marine engineering company that had spent three years designing and building the 3.3-ton protype nodule harvester called Apollo II. An imposing (also, extremely tall) Dutch engineer, Laurens de Jonge, was in charge of managing the Apollo II project – from approving the design through to testing it at sea. De Jonge's own interest in mining had begun about thirty years before, when he had helped to design a new method of extracting diamonds from shallow waters off Namibia. He had enjoyed the challenge of developing a successful new technology – one that was eventually bought by De Beers, the almost monopolistic diamond corporation. But this time, it was different, he told me. 'I wouldn't do diamond mining now', he said, 'but we do need these mineral resources.'

Over the few days I spent at sea, de Jonge's team of engineers would trial Apollo II in deep waters off the coast of Málaga. I must admit that the gulf of Málaga didn't, at first, seem to me the most obvious location for a deep-sea mining trial. It's not nearly as deep as the Pacific, and there are no valuable minerals to be found there. But for a new industry that was testing what was essentially a bespoke, multi-million-dollar robot, it was a much safer bet than sending Apollo II straight down to the deep abyss and hoping for the best. Royal IHC's plan was to test the Apollo II prototype and, if successful, to scale it up for eventual use in the Clarion-Clipperton Zone.

At full scale, Apollo II would measure 50 feet wide, gathering about 450 tons of nodules per hour; that's over 100,000 tons per fortnight. This one machine – over the course of its life – could strip nearly 4,000 square miles of seabed of its nodules. As well as removing the nodules, deep-sea mining will churn up sediment twenty-four hours a day, seven days a week for thirty years at a time, which is the duration of a proposed mining contract. Deep-sea mining comes with numerous, uncertain risks for the ocean environment, among them noise and water pollution and interference with climate regulation, but perhaps most concerning is the threat of species loss and even extinction. The Clarion-Clipperton Zone, where nodule harvesting is mostly likely to happen, is a slow, quiet environment. Sediment settles here at a rate of less than half an inch every thousand years. There's little to disturb life. So, for ocean creatures, the biggest threat comes from (of all things) the simple fact that mining disturbs the sediment of the deep sea. One of the reasons Henko and his team were collaborating with Royal IHC was to see if anything, at all, could be done to reduce this threat.

With nodule mining, the sediment comes from two sources. The first source is the mining machine itself; a harvester such as Apollo II has twin tracks, similar to those on an army tank, which allow it to crawl over the seafloor. This whips up a thick layer of fine silt as the machine travels along. But sediment also sticks to the nodules and is sucked up with them as they are collected and transported to a surface vessel through a pipe called a vertical riser. On the ship, the nodules are cleaned and any remaining sediment is then pumped back to the deep ocean through another long pipe. For reasons of cost, most contractors plan to release this waste sediment at around 3,000 feet above the ocean floor, or even closer to the sea surface: an extra 3,000 feet of piping – especially of piping that can withstand several pounds per square inch of pressure – is expensive. What little scientists do already know suggests that sediment kicked up by a nodule harvester could travel for hundreds of feet or even several miles across the ocean before resettling, burying and suffocating animals in the process. In addition, sediment raining down from above will cause problems, especially for mid-water swimmers, such as jellyfish and sea butterflies, and for commercial species – tunas, billfish and octopus – which forage there. The potential impacts are almost too complex to imagine.

On my first evening on the *Sarmiento de Gamboa*, as the sun sluggishly sank below the horizon, Henko and I sat on the ship's deck, perched between two shipping containers, and talked through the plans. Over the following days, they would set up an array of underwater cameras and sensors to monitor Apollo II's every move, to measure how much the nodule harvester squashed the seafloor and how much sediment it kicked up as it journeyed along.

On the seafloor, they had positioned six moorings, each representing a station that Apollo II would drive past. The plan read a lot like a bus timetable, but one in which connections had to be perfectly synchronised. At each station, a pair of motion-triggered sensors would measure the density of particles in the water just as Apollo II drove past, while an underwater vehicle tethered to the ship and mounted with a camera, the ROV, would video the sediment stirred up along Apollo II's tracks. Before and after the drive-by, they would sample Apollo II's 'track' (the bus route) with 'box cores' – huge white hollow tubes – to gauge the extent to which the vehicle had squashed the seafloor. The timing of the sensors was crucial. If they were triggered too early or too late, they wouldn't catch the full show, and the data would be spoiled. Another concern was interference from nearby ships: the site needed to be quiet, with as little disturbance as possible.

The design of Apollo II relied on something called the 'Coanda effect'; essentially, the pressure difference between the surrounding water and the nose of Apollo II generates lifting energy to guide the nodules into the machine. This takes up less sediment than, for instance, a suction pump. But the speed also affects how much sediment was churned up: if Apollo II moved slowly the sediment would resettle quickly and over a small area. But if Apollo II was operating at full tilt, the diffused particles would drift further in the water, taking longer to settle on the seabed; when they did eventually settle, they would cover a larger area. By monitoring the sediment intake, they could tweak the design, or the operations, to minimise the environmental impact. The reasons for doing this were at least as practical as virtuous: the more sediment that comes with the nodules, the more that has to be processed, and there's an extra cost of returning this to the ocean floor.

As Henko explained all of this, I wondered whether we could end up with slow-mo Pompeii-like scenes in the deep abyss,

3 miles underwater and several hundred miles offshore? In the deep sea, relatively large mobile creatures such as the dumbo octopus or the gummy squirrel could perhaps flap or crawl their way to safety, but what of sedentary types – creatures such as sponges and anemones? Perhaps seabed mining would, like land mining, have its own sacrifice zones – a terrifyingly fatalistic term used to describe areas that are given up and accepted as being permanently lost to all other uses. It was only in 2010, after all, that fears of an impending 'rare earth crisis' led to calls for new mines to be opened replete with new sacrifice zones, in Greenland, in the Amazon, and even on the Moon (which, like the international seafloor, is the common heritage of humankind). The impetus had been a territorial dispute between China and Japan over the Senkaku Islands in the East China Sea, leading China to threaten to block rare earth exports to Japan, which the country relies on to manufacture products such as wind turbines, hybrid cars and solar panels. If we were willing to sacrifice some of the Amazon for this, I thought, what were the odds that people would be far more disposed to sacrificing a remote, little-known part of the ocean's abyss, even if it is home to the gummy squirrel?

I asked Henko what he considered to be the most lasting damage from seabed mining. 'If you plough an untouched area on land, you change the whole ecosystem. Weeds come up and fill the void, like the poppies after the war. But there's a chance of restoration; within a decade, maybe you will have the original vegetation back', he said. 'In the ocean, if you remove the nodules, they won't be back within the next million years.' There's nothing that can be done to minimise that impact (other than abstaining from seabed mining in the first place), Henko told me, but the other impacts – such as the sediment – could be managed.

I wondered whether it bothered him to work with an industry that he believed could do irreversible damage. 'We mine on land using child labour in the Congo now. If we can mine deep-sea resources in a controlled manner, then why not?' he said, imagining, optimistically, that one might eventually replace the other.

In Congo, at least 20 per cent of the mines are operated by casual workers, many of them children who work twelve-hour shifts in deep underground shafts. Deep-sea mining is arguably one way out of this, though with valuable ores still left on land, there is no indication that large mining companies such as Rio Tinto are about to move their land-based operations offshore. Those interested in seabed mining are mostly opportunistic start-ups, looking to get a foothold in the industry. Still, Henko believed that they would do as they wished, with or without him, and that he'd rather look back and know that he'd tried to help industry develop in a less harmful way. He added, though, that he felt frustrated by the efforts to give seabed mining a green veneer, as though the industry was selling an eco-product rather than launching an extractive industry. 'It's a lie. There's nothing green about it', he said.

THE FIRST PERSON to attempt to assess the likely harm from mining the seabed was a young ecologist named Hjalmar Thiel. In 1972, when working at the University of Hamburg in Germany, he managed to secure a place on a boat with a group of prospective miners bound for the Clarion-Clipperton Zone. Thiel's own interest was in the region's largely unstudied meiofauna – tiny animals that live on and between the nodules. His travel companions were more eager to harvest the region's riches. At the time, companies from industrial nations including the US, Germany, Japan, Canada and Belgium were interested in developing the technology to harvest nodules from the CCZ, but they hadn't given any serious consideration to the potential environmental fallout.

Now retired and in his nineties, Thiel has feathery white hair, a Captain Birdseye beard and glasses that sit perched on the end of his nose. 'We had a lot of fights', remembers Thiel of that first trip to the CCZ. On another voyage, Thiel visited the Red Sea with would-be miners keen to extract valuable ores from the region's metal-rich muds. At one point, he cautioned them that if they dumped sediment at the sea surface, it could suffocate small swimmers such as plankton. 'They were nearly ready to drown me', Thiel recalled of his companions. In yet another confrontation, Thiel suggested that companies interested in mining should, at least, test the environmental impacts beforehand. He was curtly advised to do his own test. So, in 1989, he did. Thiel conceived a scheme just like Henko's, but in the Peru Basin, a part of the Pacific Ocean relatively close to the CCZ. Much like on the Apollo II trial, Thiel planned to study the impacts of a prototype nodule harvester on the seabed, and he persuaded industry to fund the test. But about a year before his planned expedition, the funders backed out, realising the scale and cost of the operation were more than they'd bargained for.

Thiel received the news at a meeting one autumn afternoon and, crestfallen, he headed back to his university office in Hamburg. As he drove through rural Germany, watching the farmers ploughing the fields, he had an epiphany. Rather than aborting the whole mission, could he make do with a simpler device to 'mine' the seafloor instead of the machine he'd been promised? In fact, perhaps he could use an oversized farming plough? Would that be so crazy? Thiel asked an agricultural machinery manufacturer to make him an oversized plough, which he named the plough harrow. In 1989, Thiel and a colleague, Gerd Schriever, carried out a simple trial that involved raking the centre of a roughly 4-square-mile plot in the Peru Basin with this odd, bespoke, implement. The experiment, called DISCOL (disturbance and recolonisation), created a plume that rained down and buried most of the study area, smothering creatures on the seafloor. Even now, more than thirty years later, neither the disturbed areas nor the nearby areas that were unploughed but affected by the shower of sediment have regained their biodiversity. All told, this suggests that the impact of seabed mining will last much, much longer than Thiel and his colleagues ever imagined. Since Thiel's experiment in the Peru Basin, there have been many attempts to advance his basic approach. There have been just as many setbacks.

Standing on the deck, I gazed in awe as Apollo II was readied for its most ambitious voyage yet. Sporting shiny aluminium caterpillar tracks against a canary-yellow and black body, Apollo II had all the gee-whiz of a space rover about to be deployed on Mars. Close to the bow, on either side of the ship's metal frame that lifted her, stood the five anxious engineers who had designed what looked like a futuristic toy. Its single, 3.3-ton structure had all but consumed the past three years of their lives. As they focused silently on the task ahead, they wondered how it would fare beneath the waves, in the fluid darkness below our sturdy, safe vessel. Apollo II dangled precariously from the winch block overhead, looking fully outsized compared to the tiny rope that suspended it just several feet above the glistening blue beneath. As it was lowered carefully into the calm waters below us, it suddenly lurched from side to side, sending a panicked look among the crew, before making a faint splash as it hit the surface. There, it bobbed around, like a diver struggling to neutralise their buoyancy, for just a bit longer than I'd expected. As the engineers watched their opus vanish beneath the waves, they breathed a collective sigh of relief.

But within a matter of hours, several problems had emerged. The first was the presence of nearby ships, which were stirring up sediment, potentially ruining the experiment. Henko was visibly stressed, and the mood on board turned glum. The second problem was more worrisome still. In the darkness below us, Apollo II was sinking into the oozing mud and needed to be retrieved, a task that involved rousing the Spanish crew from their midday siesta – an ordinarily unconscionable act but, under the circumstances, entirely necessary. Hauled onto the deck, Apollo II looking fairly beaten up – the sea had been choppy and either on its descent or at some stage on the seafloor, both of its tracks had become detached and would need repairing, as would one of its cameras. The engineers would have hours of repairs ahead of them, stretching long into the night, before trying again.

The next day, the mood on board had lightened considerably. The sea was calm and, despite the disappointment of the previous day's events, all boded well for the day's mission. Before the day's business got underway, I grabbed an opportunity to talk with Laurens de Jonge, manager of the Apollo II project. Previously brusque – or perhaps just busy – he was keen to talk. 'To those who give me flak', he said, 'I tell them that this isn't just about your car, your laptop or your iPhone. The colouring in your clothes is from zircon that is mined from mineral sands. The laces in your shoes. They are cotton; that cotton has to be braided, and you need steel to do that. Even a sock doesn't grow on a tree. You'd have to knit it, and if you didn't knit it, then it has been made by a machine, which has been mined', he said. It was a sobering point, but I couldn't help think that the prospect of mining new resources has made us lazy about reusing and recycling the minerals that have already been mined.

Surely there's potential for developing a more circular economy as an alternative to deep-sea mining? 'Urban mining' – extracting gold, silver, rare earth metals and copper in the waste generated by the disposal of our PCs, mobile phones, tablets and other electronic products – could be used to recover and recycle vast quantities of materials, reducing our reliance on other forms of mining. Currently we recycle less than 20 per cent of our e-waste – and our use, and disposal of it, is only growing. In Europe alone, around 160 million mobile phones are disposed of every year, containing the same minerals that prospective miners are targeting in the deep sea. Tackling the built-in obsolescence that has become a policy of twenty-first-century design is another way of reducing our need for new mining operations. Of course, there is a need for mining to continue – somewhere, in some form – but given the potentially irreversible destruction of deep-sea environments, I couldn't help but think we should continue to do this on land, where we've already opened mines and where there are still plentiful reserves.

While we prepared for Apollo II's second outing, Henko talked me through a new scheme for the survey. He started by showing me the digital display from the acoustic doppler current profiler (ADCP) – an instrument mounted on the bottom of the ship to measure particles in the water. It detects sediment, but also larger items such as plankton and fish. As it was morning, we could see lots of dots near the bottom of the screen. These were plankton that had spent the night hanging out at the surface and were now diligently making their way to the seabed on the return leg of their daily vertical migration. By night, these tiny swimmers would, again, be surface-bound, freeing the seafloor from any noisy interference. And with no fishing boats in sight, the entire neighbourhood would be at peace, giving them the best chance of gauging the disturbance caused by Apollo II. They would deploy Apollo II at night, and cross their fingers that the weather cooperated.

But the evening's events didn't quite work out. As Apollo II was being lowered, a 6-foot swell hit the ship, jerking the mining machine up and down in the water as the cable slackened and tensed with every roll and pitch of the vessel. On the video camera, Apollo II looked shaken, to the point where the crew worried they might lose the machine entirely. And so up it came again, and this time, the extent of the damage was much more severe. Disappointed, the team decided to call time on Apollo II's mission to the deep and return to shore where they would consider their next steps. The whole experience made me appreciate how difficult it is to gauge the impact of a nascent industry with new machinery, operating in an environment that is largely unknown, and virtually inaccessible.

Since the Apollo II expedition, there have been other similar trials, and some rather more successful. In May 2021, a prototype mining machine named Patania II was successfully deployed in the Clarion-Clipperton Zone (CCZ). Developed by a Belgian company named Global Sea Mineral Resources (GSR), the 30-ton robot sucked up nodules from the seafloor while scientists on a nearby research vessel monitored the impacts, in an attempt to gauge the potential ecological harm. More recently, in November 2022, The Metals Company – a Canadian start-up – went further still, not only collecting nodules from the CCZ but transporting them through a vertical riser to the sea surface, proving that the company has an entirely operational system for harvesting ore-rich rocks from the Pacific abyss. As part of the trial, more than fifty subsea sensors and moorings were deployed at the site to monitor the impacts. But even now, with all of this research (and much of it funded by industry), we're only scratching the surface of understanding the ocean's abyss and the implications – for our planet – of allowing it to be mined.

In March 2022, Diva Amon – the Trinidadian biologist who I first visited at London's Natural History Museum – led a comprehensive review, with thirty other scientists, of the questions that still need answering to gauge the potential harm. They looked at various categories of information you'd need for a thorough assessment: what we know about the environment – not only the animals that live there, but how they vary over time and space and their relationships to one another – what we know of the impacts of mining such as noise and light pollution, plumes, the release of toxic metals, and what, if any, management plan is in place: if it all goes horribly wrong, how do we mitigate the impacts? According to their study, even now, there's enough information on just 5 per cent of the CCZ. Filling in the gaps would be a 'monumental task' said the authors. It would 'likely take decades' wrote Amon on Twitter.

In the absence of this information, many fear the industry is collecting environmental data as a formality only, and that regardless of what ecological harm deep-sea mining might do, it will go ahead. Their fears seem warranted: in 2021, The Metals Company stated its intention to begin mining as early as 2024, even in the absence of environmental regulations. One might think that position untenable, illegal even. Only on the high seas, it's never that simple.

FOR AN AGENCY that has jurisdiction over half of our planet, the International Seabed Authority (ISA) has, historically, attracted surprisingly little public scrutiny. Established as an independent body by the UN in 1994, the ISA occupies a large, but otherwise unspectacular, building in Kingston, Jamaica, from where its fifty staff try to fulfil an unusual, and challenging, mandate: to protect the international seabed and to meanwhile generate and distribute profits from mining the international seabed. In other words, the ISA has been hopelessly conflicted from its inception. As one legal expert put it to me: 'The ISA is both the poacher and the gamekeeper.' When the UN established the ISA, it wanted to honour a request by developing nations that they would benefit from the exploitation of international seabed resources. One way they hoped to achieve that was by allowing the regulator to also become an operator, an arrangement that has been described as 'a unique undertaking in the history of international cooperation'. The idea was that the ISA would establish a commercial mining arm, called the Enterprise, which, like all deep-sea mining operators, would partner with a sponsoring state, but whose profits would be distributed between developing nations. Early iterations of this idea had the Enterprise enjoying a monopoly on resource extraction from the deep sea. These days, the vision for the Enterprise, which is still being wrangled over, is that it will take over some plots of subsea real estate for mining – the profits of which will be distributed to developing nations – as well as having a role in transporting, processing and marketing minerals that are mined from the deep sea by other commercial operators. Either way, the ISA will be akin to a state-owned mining corporation regulating its own activities.

Even now, before any extraction has begun, the arrangement has caused problems. Many conservationists worry that the ISA, with its conflicted role, cannot be trusted to serve one of its core missions: protecting the deep sea. As evidence, conservationists point to the fact that ISA has approved every application for exploratory mining, which now number thirty contracts. Of these, seventeen are for sites in the Clarion-Clipperton Zone, and others distributed throughout the high seas. For the most part, 'contractors' are national governments (in the case of India), or research institutes (in the case of France), or commercial companies (e.g., The Metals Company), that have teamed up with a sponsoring state (in the case of The Metals Company, the sponsoring states are Nauru, Tonga and Kiribati) to explore a claim area on the high seas, with the assumption that it will eventually mine this area commercially. One especially contentious case has been the contract awarded to Poland in 2018 to explore a vast abyssal site in the mid-Atlantic, next to and partly in the site of Lost City's hydrothermal field. One of the most extreme environments ever found, Lost City has been earmarked for World Heritage status.

All told, the implication is that ISA, an agency with responsibility for ensuring that mining causes no 'serious harm', is willing to allow excavation to proceed with little regard for the impacts. In response to its critics, the ISA has pointed out that it has been proactive in establishing an environmental management plan ahead of mining. This includes setting aside nine areas of particular environmental interest within the CCZ. The intention is to keep these areas – about 30 per cent of the region – free of mining to protect biodiversity. What's more, contractors are required to conduct baseline ecological surveys of their claim areas, and they can carry out tests to understand the environmental impacts of their machines. But scientists say the system has not worked well in practice, in part because the baseline requirements are so weak, and there are no legal obligations specifying how contractors should test the environmental impacts of their mining machines.

More worrying still, however, is the fact that ISA could, in theory, allow commercial mining to proceed without having regulations in place – rules of the sort that would hold industry accountable and legally responsible for damage to the planet. Under ISA rules, sponsoring states have a right to invoke something called the 'trigger', which stipulates that they can begin deep-sea mining two years after they submit their plan of work, regardless of whether there is a mining code (there was not at the time of writing), with environmental regulations, in place.

In June 2021, The Metals Company, in partnership with the small island state of Nauru, invoked this arcane rule, announcing their intention to begin harvesting nodules from the Pacific seafloor as early as 2024. At the helm of The Metals Company is Gerard Barron, a self-styled climate crusader who claims he will transition the world off fossil fuels and help create a circular economy, even while his critics say he's about to wreck one of the most pristine environments on Earth. Barron, a native Australian whose background is as a 'strategic investor', has been chief among those pushing the ISA to green-light mining. According to The Metals Company's own estimates, if it manages to extract the vast stashes of cobalt, copper and nickel that is allegedly held in its three claim areas, it could power a staggering 280 million electric vehicles, comparable to the entire existing US car fleet. Its estimated earnings would be $31 billion over twenty-five years.

In February 2023, just weeks after The Metals Company demonstrated its ability to haul commercially viable quantities of manganese nodules from the Pacific abyss, I met Barron in London for a coffee. My request for interview had been answered, somewhat surprisingly, in the affirmative, and for an hour or more, Barron – presumably buoyed up by recent events – spoke openly, and effervescently, of his plans for The Metals Company and of his reactions to his critics. He told me that some of his detractors were trying to 'make mountains out of molehills' by, for instance, alleging that the company had released sediment in an unsanctioned and uncontrolled manner from the sea surface into the ocean during their recent trials. The seriousness of the incident, which was posted in a video online, seemed fairly nebulous. But Barron then went on to make some further claims about the potential for deep-sea mining to have a regenerative effect on areas. 'What we're finding now is that actually, when you do disturb an area, it creates much more attraction ... for organisms to come back. Once we start to collect nodules, the area starts to rehabilitate pretty well straight away', he said, adding, 'But we've got to keep in mind that we're talking about a very confined space as well. We're talking about a very small footprint.' The nodules, of course, are removed in perpetuity. As for mining's restorative effects, I have yet to see evidence of this.

Barron first came to seabed mining as an investor in a Canadian start-up called Nautilus Minerals, which – until a few years ago – was hailed as the world's first commercially viable deep-sea mining company. Their flagship project, Solwara 1 (Solwara is pidgin for 'saltwater'), aimed to excavate minerals from hydrothermal vents in the Bismarck Sea, at a site owned by Papua New Guinea. To that end, the company successfully developed and tested three monster mining machines each the size of a combine harvester, which quickly came to symbolise the havoc industry was about to wreak on the deep ocean floor. But about halfway through construction of *New Era* – the mother ship designed to take the ores for processing – costs ran over. Amid the negative publicity, investors became wary, the coffers began to empty and, by early 2019, the company had started to liquidate. By April of that year, the company assets – including *New Era* – were being auctioned by PricewaterhouseCoopers.

Yet prospective miners appeared undeterred by Nautilus' unravelling. A few weeks after its liquidation, back in 2019, I found myself at a deep-sea mining summit in London's Merchant Taylors' Hall, a regal, imposing building on Threadneedle Street, with high stone medieval walls (that survived both the Great Fire of London and the Blitz of the Second World War) and an interior adorned with artefacts and oddities from various centuries. The marketing material for the event had attracted my attention, describing deep-sea mining as 'gold rush' and a 'gamechanger' that, after years of false starts, was close to a breakthrough. Over a couple of days spent sitting in an oak-panelled room below ornate chandeliers, my feet resting on deep-piled red carpet, I listened to talks from prospective miners and their suppliers. I was keen to hear their perspective, given the timing. One of the keynote speakers was Govinder Singh Chopra, director of SeaTech Solutions, the company that designed Nautilus' *New Era* mother ship. Chopra portrayed a healthy nascent industry suffering a few inevitable bumps in the road, with new developments happening in all directions, from improvements in underwater communications to plans for subsea power stations. I sat enthralled and horrified, imagining this underwater world that was developing, all centred on what would eventually be a mining town, 3 miles below the waves. It was captivatingly futuristic, if deeply, deeply troubling. 'Environmentalists have more or less accepted that a little part of the ocean will be affected by this', said Chopra. 'Nothing much is now holding us back from exploitation.'

As for Barron, he got out before Nautilus went bust, and recast himself as a 'green' miner, CEO of DeepGreen Metals, which since rebranded itself as The Metals Company. Again, the partnership was with a Pacific Island state, not Papua New Guinea, but for The Metals Company, the partners are Nauru, Tonga and Kiribati. On the face of it, it seems odd that nations, such as Papua New Guinea and Nauru, that have been so damaged by land mining, would buy into this. Papua New Guinea is a nation plagued by mining-related disasters. Most famous perhaps is the disaster at Ok Tedi, where discharges from a copper and gold mine have spoiled over 60,000 miles of pristine river and displaced 50,000 people from their homes and often livelihoods. There's also the legacy left by Rio Tinto, the British-Australian mining giant, which operated one of the world's largest open-pit copper mines in Papua New Guinea's Bougainville from 1972 to 1989, but was reluctant to pay for the clean-up of nearby waterways and contaminated farmland. Surely these nations, of all, would see the downsides of this extractive industry? Cash-strapped and on the front line of climate change, many, however, are desperate to find new sources of income. Indeed, Papua New Guinea was an early supporter of seabed mining, with the Solwara 1 project long seen as the frontrunner in commercialising deep-water extraction.

Now Nauru has taken Papua New Guinea's place. And Nauru is about as troubled as a nation gets – financially, it is nearly flat broke, and its recent history has been one of corruption, incompetence and environmental destruction from land-based phosphate mining. The world's third smallest nation and smallest island nation, Nauru was once among the world's wealthiest. In the 1970s, it was raking in the modern-day equivalent of $2.5 billion per year – easily enough to keep its inhabitants afloat for generations – from a seemingly endless supply of phosphate-enriched fertiliser deposited over millennia by visiting seagulls.

But phosphate mining by the UK and Australia has left 80 per cent of the tiny island uninhabitable: by removing large layers of Earth to access the minerals beneath, it has left the land barren, infertile and incapable of supporting plant life. What's more, the profits of this environmental catastrophe have been squandered in a series of bizarre investments by inept officials. Most noteworthy was the financing of a flop West End musical – co-written by one of Nauru's financial directors, Duke Minks – about a torrid love affair between Leonardo da Vinci and the Mona Lisa. Met with dismal reviews, the project lost the citizens of Nauru $7 million. Academics now consider the island a failed state, and so the prospect of regaining some of its former wealth through seabed mining – especially in international waters, far from the island itself – is an attractive one.

If irreversible environmental damage is the obvious case against deep-sea mining, then helping poor nations such as Nauru has been offered up as the counter-argument. Under current ISA rules, when a developed nation applies to mine the international seabed, the proposed claim area must be big enough to support two operations. The claim area – which doesn't have to be contiguous – is spilt into two areas of equal estimated value, with one half given to the applicant and the other set aside as 'reserved areas' to be explored by a developing country. Developing nations can partner with a contractor to explore the mineral wealth of their half and, so far, Kiribati, Cook Islands, Tonga and Nauru have partnered up with industry to do that. Even for those without any direct involvement in mining, however, there's the promise of wealth through 'benefit sharing' arrangements, simply because minerals sourced in international waters are the 'common heritage of mankind' and should be distributed among all ISA member states, which are taken to represent humanity. The question is whether any of this could ever be enough to offset the environmental damage. That's a value judgement, of course, but it's at least worth taking a look at the potential profits to be made.

An analysis from researchers at the Massachusetts Institute of Technology in March 2019 found that based on ISA's recommendations at the time, each thirty-year mining contract would be worth around $10 billion to the contractor, $4 billion to the sponsoring state and around $490 million in 'common heritage' value. According to environmental lawyer Duncan Currie of Globelaw, that works out at about 8–9 cents per man, woman and child in the world. Assuming there were thirty contracts (based on the exploration contracts that currently exist), the entire value of seabed mining to humanity would work out at $2.50–3 per person totalled over the thirty-year lifespan of a contract. In reality, of course, we wouldn't get this money back as a tax refund; one option is that it would be accrued over time in an environmental fund of some sort that would invest in advancing our knowledge of, and competence in exploring, the deep sea. But even at this level, there's an assumption that the costs of environmental remediation, if needed, would come out of the common heritage portion of the profits, rather than being paid for by the contractor. For the moment, none of this has been agreed and the profit-sharing arrangement remains under discussion.

MEANWHILE, LACKING INFORMATION about deep-sea environments and the harm that could ensue from mining, researchers have been left unsure as to how to even define the risks. 'What is serious harm? There are some clear red lines, but there's no definitive answer to that question yet', one ecologist told me. 'We understand that global extinction is serious harm and we know that interference in carbon sequestration is serious harm. Scientists know that mining will cause local extinction of species in the CCZ, but are we talking about the extinction of species across the CCZ or just in the mined area? It is complicated', he said. 'For most of the animals in the direct vicinity, mining will be lethal', said Henko. 'It will wipe out most of the large animals and everything that's attached to the nodules. That's a given, I would say.' But the impacts of mining in the CCZ will clearly have a much further reach. Just how far is unknown. One computer-modelling study found that the sediment could take up to ten times longer to resettle than is currently assumed, meaning it will probably travel further in the water column. And some researchers say that even trace amounts of sediment stirred up by the mining operations could smother seafloor life far away. 'We're only starting to see how far the plume reaches and we're still very far from knowing what the effect will be', said Henko.

Around the time I joined Henko at sea, a growing chorus of voices started to call for a moratorium on seabed mining. When an estimated 53,000 gallons of toxic red slurry spilled from a nickel mine in Madang, on Papua New Guinea's east coast, local community groups launched a campaign against Solwara 1, demanding a ten-year ban on extracting ores from the seabed. The rationale for a decade-long hiatus, which had long been championed by conservationists, was that it would give more time to explore marine life – and what needs protecting – while also figuring out the potential damage from seabed mining. That same month, the then prime minister of Fiji, Frank Bainimarama, sent a communique asking other Pacific Island nations to support the moratorium. The prime minister of Vanuatu, Charlot Salwai, immediately signed on. By September, the Papua New Guinea government – initially a staunch supporter of seabed mining – had backed it. By early 2021, several multinationals that rely on rare metals, including BMW, Volvo, Samsung and Google, had lent their voices to the call. The list keeps growing, and now includes French car maker Renault and US car maker Rivian. In June 2022, French president Emmanuel Macron called for a ban on high-seas mining and for new laws that would protect these fragile ecosystems. By early 2023 many nations had joined France, and Canada – where The Metals Company resides – had expressed concern about deep-sea mining, asking for more assessment of the impacts, both in national and international waters.

On hearing the news, I thought back to my time on the *Sarmiento de Gamboa* and on the seeming incongruencies within the mission – the engineers' pride in Apollo II and their unwavering belief that they were helping to develop a more ethical, less destructive mining industry contrasted with the growing view among scientists that this will permanently damage Earth's last remaining wilderness. At one point, Henko had offered an explanation. 'Many of us are so used to the alteration of the Earth's landscape that engineers just possibly don't see mining the deep sea as being that substantially different', he said. 'We've created entirely man-made landscapes on Earth – we've altered the path of rivers and we've removed the tops of mountains. We just haven't done this before in the deep ocean.'

5

THE INTERVENTIONISTS

IT WAS A WET, blustery day in July 2012 when the fishing vessel *Ocean Pearl* set sail from Victoria on Vancouver Island and headed out into the cold, deep waters of the Pacific Ocean. The crew numbered just twelve, and their destination was the Haida eddies. This small swirling mass of ocean currents lies just beyond Canada's territorial waters, hugging the coast of Haida Gwaii, a remote Canadian archipelago positioned northwest of Vancouver. At the boat's helm was Russ George, a man who liked to go by the name of 'Greenfinger'. He claimed he was on a mission to save the world, and to make some money on the side. In a chequered career, George had tried various routes to achieve this aim, involving himself in the development of cold fusion, sablefish farming, and shiitake mushroom ranching. None of his ventures had, so far, brought him wealth or prestige.

This trip to the Haida eddies was his best chance yet. His plan was to seed the open ocean with iron sulphate – the same sort of iron dust used in common fertilisers – in the hope of stimulating a bloom of plankton large enough to soak up millions of tons of carbon from the atmosphere, drawing it down into the deep sea, and simultaneously replenishing local salmon stocks, by boosting the zooplankton and fish larvae on which they feed. That same month, scientists reported, in *Nature*, the first evidence that fertilising the ocean with iron could store carbon in the deep sea for centuries or even millennia. In practice, what happens is that given enough nutrients, iron being one, plankton grow rapidly on the sea surface in spring and summer, absorbing atmospheric carbon as they grow. Soon after, however, plankton that haven't been eaten die and fall in huge gelatinous clumps to the seabed, bringing this carbon with them. So, if George's scheme worked as planned, the American could claim the honour of having restored the declining fisheries of Haida Gwaii, while also having stored vast quantities of carbon on the seabed, keeping it locked out of the atmosphere for centuries, and perhaps even millennia. Any carbon buried at sea could then, in theory, be sold as credits on the international carbon market, at which point George and the residents of Haida Gwaii would reap financial benefits from the scheme.

As it turns out, however, the Haida Gwaii experiment would also be the largest attempt ever to alter the Earth's climate. By 2012, scientists had dreamed up all sorts of different schemes to stall some of the warming expected from human-induced climate change. These ranged from the pragmatic, such as painting roofs white to deflect sunlight and planting forests to soak up carbon, to the seemingly eccentric, such as deploying enormous sunshields into space. Like all climate intervention schemes, ocean fertilisation has always been understood to have risks. Perhaps it would end up nourishing toxic plankton blooms that could spoil fisheries – or the blooms might grow so big and spread so widely that they would starve the ocean of sunlight, nutrients and oxygen.

However, on that July afternoon in 2012, excitement was running high. In the local town of Old Massett, one of the largest Indigenous communities on the Haida Gwaii archipelago, residents were hopeful that George's experiment would work. Over the previous decade, the once plentiful sockeye salmon population had seen a precipitous decline. On Haida Gwaii, where most residents had once relied on fishing for their livelihood, unemployment had reached a record 60 per cent. A year earlier, George had approached Haida Gwaii's council members to win their support, and financial backing, for the project. Desperate to see a return of their bounty, and also hoping to participate in the carbon credit gold rush that could ensue should George be successful, the villagers had voted two to one in favour of George's plan, committing $2.5 million to his company, the Haida Salmon Restoration Corporation (HSRC).

After hours of sailing through driving rain and high winds, George's team were on site. There, they mixed approximately 100 tons of iron dust with seawater and, using a hose, pumped the mixture straight into the swirling currents of the Haida eddies. The team watched as the waters turned a muddy red, as if a Sahelian storm had passed through, dumping its load right into this one tiny patch of otherwise radiant blue ocean. About a month later, the crew returned to the site, and topped up their initial iron injection with another 20 tons. All in all, roughly 130 tons of iron sulphate had been dumped at sea. Within days of the second discharge, George's dream was coming to life before his own eyes. Over nearly 4,000 square miles, he could see 'a verdant emerald sea lush with the growth of 100 million tons of plankton'. The bearded, unkempt American told of large numbers of fin, sperm and sei whales, orcas, dolphins, tuna, squid and seabirds visiting the site, some of which were rarely found in the region.

George's plan was going swimmingly. Now all he needed to do was wait until the plankton died and sank, bringing their precious load of carbon to the very depths of the ocean, where – in theory at least – it should stay locked up for centuries, and out of the atmosphere. But there was a problem, and a serious one at that. The prospective planet-saver had disregarded not one, but two major international conventions. This was not George's first attempt to bury carbon at sea, and make a few quick bucks on the side. In 2007, while heading a company called Planktos Corp., he had planned a very similar experiment on the high seas just beyond Ecuador's Galápagos Islands, and another near Spain's Canary Islands in the Atlantic. Environment organisations Greenpeace and the International Union for Conservation of Nature got wind of his plans and raised the alarm, calling on lawmakers to intervene. Spanish and Ecuadorian ports then refused him access, and the US Environmental Protection Agency warned him against flying a US flag for his Galápagos trip. He backed off. In the months that followed, two major bodies – the UN Convention on Biological Diversity (CBD) and the International Maritime Organization (IMO) – called for a moratorium on ocean fertilisation experiments, other than those being done for legitimate scientific research. These statements should have been enough to warn George off his Haida Gwaii experiment, but neither declaration was legally binding. The rules are voluntary and, in the case of the IMO's ban on commercial iron fertilisation, only the UK had signed up.

The irony is that while Russ George's Haida Gwaii project was unsanctioned, ill-conceived and premature, it was grounded in science. The fact is that the ocean has an enormous capacity to soak up carbon and if it could absorb a bit more, it would buy us crucial time in dealing with the source of the climate problem. While seeding the ocean with iron is one way of doing this, it's just one of many practices that could physically remove CO_2 from the atmosphere, lessening the harm caused by greenhouse gas pollution. Since the Industrial Revolution, humans have pumped an extraordinary amount of CO_2 pollution into the atmosphere. Each year, around 44 billion tons of carbon dioxide are sent straight into the sky, where they alter the atmospheric chemistry, trap heat and change our planet's climate. As a result, atmospheric concentrations of CO_2 are now 48 per cent above pre-industrial levels and while we know how to slow this upward trend – keeping fossil fuels in the ground – we're doing quite the opposite.

In all of this, the ocean in general, and the deep sea in particular, has shielded us from much of the potential damage of heat-trapping CO_2, by soaking up roughly a quarter of the emissions that we've pumped into the air. It's also absorbed 90 per cent of the extra heat we've generated in the past fifty years. To put that in perspective, if all of that heat had entered the atmosphere instead of the ocean, the global average temperature would have jumped by roughly 99°F, according to one estimate. The ocean has been able to soak all of this up in part because it is 280 times larger than the atmosphere. It is also denser than air, holding up to four times more heat per unit area.

But in providing this vital carbon- and heat-storage service, the ocean has changed dramatically. For a start, seawater is now more acidic than at any point in the past 2 million years. In the relatively short period over which we've burned fossil fuels, ocean acidity has increased by about 30 per cent, a rate of change around 100 times faster than any experienced during the past 55 million years. While the ocean's increasing acidity is invisible to us, it's been disastrous for the ocean's 'calcifiers' – those creatures, such as corals, crabs and scallops, that construct their skeletons and shells from calcium carbonate – a compound that dissolves in acid. By some projections, ocean acidity could double in the future, with unknown consequences for ocean life as we know it.

At the same time, the ocean is now reaching its limit as a heat repository. Already, the seas have warmed to depths below 3,000 feet, changing the conditions for marine life: in response, fish and other marine animals are shifting their locations, and some, such as octopus and squid, are finding it harder to grow. Fuelled in part by climate change, the ocean now has at least 700 dead zones, areas so low in oxygen that they struggle to support life. Added to that are large-scale structural changes, including the slowing of major ocean currents such as the Gulf Stream. Scientists don't yet know the effect that storing more CO_2 in the ocean will have for the deep sea and its marine life, nor do they fully understand the consequences of any proposed scheme to fix the climate. And so, following Haida Gwaii, the idea of climate intervention became taboo, seen by some as lazy and risky, set against the much harder task of transitioning to a zero-carbon economy. George's experiment was cast as the work of a huckster.

For George, the whole episode ended messily: his business partners at the Haida Salmon Restoration Corporation claimed he had lied about his expertise, and in doing so convinced them to proceed with a dubious experiment. The Canadian government seized his data for investigation. George has denied the allegations and filed a lawsuit for defamation against the HSRC. According to George, who now lives in the English countryside, it has all been terribly unfair. 'The story was that I was an independent and rogue geoengineer', said George in 2019. He is adamant that given enough time, he could have proved the merit of his work.

But even now, over a decade later, there are significant challenges in proving that iron fertilisation works and little robust evidence. Without knowing whether the carbon is removed from the atmosphere, and stored safely, it's impossible to sell carbon credits from such a scheme on the international carbon market. With so many reasons to rail against these emerging technologies, in the aftermath of the Haida Gwaii experiment, coffers dried up, field trials stopped, and many legitimate climate researchers looked to distance themselves from the controversial practice. According to ocean chemist Greg Rau who works on climate intervention technologies, 'The experiment by Russ George set the whole field back at least a decade.'

YET AS RESEARCH slowed, emissions continued to rise. Behind the scenes, politicians started to panic. In Paris in 2015, in the midst of brokering an international deal on emissions reductions, the UN invited its body of climate experts, the Intergovernmental Panel on Climate Change (IPCC), to investigate the impacts of climate warming of 1.5°C (2.7°F) above the pre-industrial global average, a level beyond which there is a risk of catastrophic impacts. The IPCC had until then largely focused its attention on efforts needed to avoid 2°C (3.6°F) of warming, but by 2015, it was becoming obvious that even that wouldn't be enough to keep everyone safe. Their conclusion, made public in 2018, was that the pledges made in Paris were grossly insufficient and that without physically removing CO_2 from the atmosphere, there's little, if any, chance of avoiding a global temperature rise of 1.5°C (2.7°F).

In practice, the scientists used high-powered computers to model future climate using a range of scenarios; different scenarios play out depending on how much energy we use, the extent to which there is universal access to renewable energy sources, for instance, and whether we can physically draw down CO_2 from the atmosphere. For every scenario consistent with keeping warming below or at 1.5°C (2.7°F), we would need to dramatically reduce emissions by 2030. Otherwise, we will likely see the loss of 90 per cent of the world's coral reefs, the prevalence of heat extremes across the globe, the start of entirely ice-free summers in the Arctic, and the loss of low-lying islands such as the Maldives as habitable nations. From mid-century onwards we will need to remove around 1 billion tons of carbon from the atmosphere each year to avoid these outcomes. And even if we zero our emissions now, the CO_2 we've already pumped into the atmosphere, all 1.7 trillion tons of it, will continue to warm our climate. It's this 'legacy CO_2' that carbon capture can deal with. Of course, we could overshoot 1.5°C (2.7°F) and then try to claw our way back at some later stage, but that won't be without consequence. Far from scoffing at the mere mention of climate intervention, the scientific community – and the policy-makers they were advising – had entered a new phase, a more realistic appraisal of our current plight. When the IPCC started to consider, in earnest, the options for avoiding dangerous climate change, it became obvious that some of the schemes we'll be forced to consider may be radical, with unintended consequences.

The IPCC didn't consider the most radical of climate intervention options: to block incoming sunlight or reflect more of it back to space. Instead, it focused solely on carbon dioxide removal (CDR) technologies – in other words, methods to clean up the existing mess and put our polluting gases into the geological equivalent of landfill. Their reasoning was the fact that removing CO_2 tackles the climate problem at its root, rather than simply alleviating the symptoms; a related concern is the fact that if solar intervention was deployed, and then abruptly terminated, temperatures could skyrocket, making parts of the planet largely uninhabitable. Carbon dioxide removal was a softer entry point.

Oddly, the IPCC focused its attention on how carbon removal might be achieved on land. The option seen as most favourable was BECCS, an acronym for bioenergy with carbon capture and storage. In this scheme, fast-growing plants such as perennial grasses are used as a fuel source, with the CO_2 produced in their combustion then captured in a flue and buried deep underground, possibly using existing infrastructure such as decommissioned oil wells. The upside of BECCS is that some of the technology already exists but, in practical terms, BECCS requires so much land that it competes with space to grow crops, posing a risk to food security. 'The emphasis for a long time was on land – on trees and plants and soils – without really thinking about the ocean, and I think that was simply a human bias in familiarity', said Greg Rau, the ocean chemist who researches carbon capture and storage through his Ottawa-based company Planetary Technologies. 'I and others baulked at that because, you know, it's a big Earth and most of it isn't land, it's mostly ocean. So why restrict ourselves to land?' The focus on land, he said, was also part of a general mindset that the oceans are to be preserved. The 'hands off – we need to protect the ocean and not do anything with it' ideology now seems ill-conceived, said Rau, as our efforts to reduce emissions have failed.

Anyone who has cared to listen has been wracked by the realisation that we can't afford to dismiss using tech to fix the climate. And so, the mood has started to shift. Since the release of IPCC's influential report, that slow dawning has become a movement, informed by scientists, energised by advocates and increasingly funded by billionaire philanthropists. All agree that it's of the utmost urgency that we figure out how, exactly, we can cool the climate, or at least avoid accelerated warming in the future. With the realisation that land-based solutions for climate intervention may come at a considerable cost, experts are turning their attention to the high seas. 'This is where it makes most sense, because there's no conflict of interest with any other issues', said Ulf Riebesell, an oceanographer in Kiel, Germany. 'These areas are largely unused at the moment and they also make up 50 per cent of the planet's surface.' And so, a decade after Russ George's divisive experiment, it finally looks as though someone will have another crack at the whip.

IN JANUARY 2021, billionaire tech entrepreneur Elon Musk announced that he was offering $100 million in prize money for the best technologies for capturing, and storing, atmospheric CO_2. The news was received as welcome respite from the gloom of the global Covid pandemic (which, by then, had reached 100 million cases and rising). Within hours, Twitter was buzzing with speculation as to who could snag the prize money and for which interventions. Many replied, sardonically, with a single word: trees. Some clearly thought that we've all become too wonkish to see the obvious solution before our eyes. (Unfortunately, climate change is already increasing the prevalence of droughts, wildfires and pests, all of which means that trees are unlikely to remove and store as much carbon as we'd once hoped.)

Others, concerned with the impracticality of turning vast swaths of land over to climate-fixing tech, suggested storing carbon at sea. One idea that kept popping up on Twitter in response to Musk's $100 million cash prize was 'enhanced weathering', a method that involves the accelerated weathering of silicate rocks – such as olivine – to boost undersea carbon storage. Currently being investigated by various non-profits and academics, including Greg Rau, the most public proponent to date has been Eric Matzner, an ex-Silicon Valley tech entrepreneur turned climate intervention enthusiast. Matzner's hope is to solve the climate crisis by adding approximately 1 trillion tons of crushed olivine rock to the ocean. The idea is this: the crushed rock, with its large surface area, reacts with the CO_2 in seawater to form bicarbonates; eventually these bicarbonates become incorporated into the shells and skeletons of sea creatures. Once they die, the carbon is carried down to the deep sea, where it is stored for centuries.

Matzner previously found fame as a biohacker and futurist. In 2014, he launched Nootroo, a San Francisco–based start-up that sold cognitive-enhancement pills or 'nootropics' (from the Greek 'to bend or shape the mind') in gold and silver formulas, to be taken on alternating days. The gold pills contain the memory aid noopept, while the silver ones contain something called phenylpiracetam, a substance that is alleged to boost the stamina of cosmonauts. During his years at the helm of Nootroo, Matzner became a poster child for the nootropics industry, capable of typing 150 words per minute, blurting out words at a speed faster than your average auctioneer, and winning calmness contests as testimony to the power of his daily meditation practice. Whether any of this was due to the smart drugs, or just Matzner's general gusto is unclear – to date, the efficacy of nootropics is unproven, in part because they're marketed as supplements and, as such, free of the regulatory requirement of FDA approval.

In 2017, however, Matzner changed direction and took a break from Nootroo, joining a team competing to build the Hyperloop, a tube-based transportation system, for Elon Musk, and simultaneously launching a think tank called Climitigation. Intended as a cogent response to global warming, its purpose was to investigate technologies for reversing existing human-inflicted damage to the Earth's climate system. Matzner soon realised that there were few real-world studies into any type of climate intervention and, after considering the options, launched Project Vesta in 2018 with David Sneider, another ex-Silicon Valley entrepreneur, to see whether enhanced weathering could work. 'This technology was sitting squarely in the Valley of Death', he told me, 'but it has potential – it's relatively cheap and it's permanent. It just needs to reach maturity.' The beauty of enhanced weathering is that it not only removes CO_2 from the atmosphere, but it also counters ocean acidification, which is threatening the world's fragile coral reefs. Matzner's plan is to deploy a trillion tons of rock to the ocean at coastal locations, dotted throughout the globe. For the time being, the project is not for profit, and the data, once collected, will be accessible online. While Matzner's transition from cognitive enhancement to climate intervention might seem arbitrary, he explained that they are both part of his overall vision for a longer, better life on a healthy planet. 'Part of my grander vision of the future and of human optimisation is about trying to live longer, beyond 100', he said. In June 2021, Project Vesta, which has since spun out from Climitigation, secured an undisclosed location in the northern Caribbean for its first pilot test with the eventual aim of global deployment. It sounds promising but, as with many of the climate interventionist ideas, there are considerable risks.

The addition of olivine sand could, for example, turn the world's beaches green. Or, as independent experts have highlighted, it could do the same to the ocean, if silicon-fixing plankton such as diatoms and nitrogen-fixing plankton such as cyanobacteria thrive in an olivine-enriched sea. Alternatively, calcifiers could multiply on the added calcium carbonate, which they use to construct their shells and skeletons, turning the open ocean white. Either way, enhanced weathering will impact the ocean, and marine life, far offshore in ways that are yet unknown.

Project Vesta are the not the only ones looking into this. In September 2019, California non-profit Oceankind hosted a conference of ocean chemists in Half Moon Bay, California, to discuss the costs and feasibility of deploying approximately 5 billion tons of silicate rock, twice the amount used in global cement production, in the ocean each year, using a fleet of ships traversing the high seas. Risks aside, the energy requirements involved in mining this amount of material, let alone deploying it worldwide, are astounding. It's one thing to test climate intervention technologies, such as enhanced weathering, for potential negative outcomes, but who is assessing the overall environmental, and climate impact, of a project that involves such large-scale extraction from the outset? No one, it seems.

Meanwhile, other ideas abound. According to some experts, there are at least twenty-six different ways that we could use the ocean to fight climate change. Almost all involve storing large amounts of CO_2 in the deep sea, but of the many ideas for how we would do that, just six, including enhanced weathering, are worthy of further investigation, according to a preeminent body of American experts. In December 2021, the US National Academies of Science, Engineering, and Medicine (NASEM) released a long-awaited appraisal of these viable ocean carbon storage options. NASEM's analysis, funded by US non-profit ClimateWorks, looked at how well-developed, feasible, risky and costly each approach might be. While many of these innovations sound like the stuff of science-fiction, they are serious suggestions and ones that – faced with runaway climate change and on a trajectory to increase, rather than reduce, our emissions – we must now consider.

The first option that NASEM looked at (somewhat annoyingly, I imagine, for Russ George) was ocean fertilisation. Scientists have had more than two dozen attempts at this in the real world. While most have failed to prove any long-term storage of CO_2, one trial, in the Southern Ocean, was successful. Just like George, the researchers doused a small patch of the ocean with iron sulphate, topped it up two weeks later, and watched as it bloomed. Measuring approximately 65 square miles on day one, by day nineteen, the fertilised patch had grown to 308 square miles. Four weeks after fertilisation, there was a plankton bloom big enough to be visible from space. As the plankton died off, they clumped and fell to the seafloor, at least half of them reaching below 3,300 feet and a substantial portion touching the deep seabed, where the carbon, in theory, would stay locked up for centuries. Scientific experiments aside, we do also have evidence that ash deposited from volcanic eruptions has boosted iron levels in coastal waters, leading to local plankton blooms, but whether or not this results in carbon being transferred to the deep sea remains unclear.

In a similar vein to ocean fertilisation, seaweed cultivation on a massive scale is also being investigated as a way of enhancing ocean life to lessen the blow of our greenhouse gas emissions. The past decade has seen seaweed production double – with an estimated value of approximately $59 billion in 2019 – as interest in seaweed for food, as a renewable material and as a carbon sink has grown. For the purpose of ameliorating climate change, the question is whether cultivating seaweeds – also known as macroalgae – at a large scale will capture and store a meaningful amount of carbon. The issue is not just growing the seaweed, but sequestering the CO_2; if the seaweed isn't harvested and stored at the right time, the captured CO_2 will make its way back into the atmosphere. So, scientists are proposing that we sink it in the deep sea, burying it in deep ocean sediments. The logistics of this are tricky, and what little research exists casts doubt on whether it would even work. One study in the North Pacific Ocean looked at the carbon benefits of placing 220-pound bales of giant kelp on the seafloor at 5,500-foot depth; after six months, larger and more diverse fish were found near the seaweed bales, but there was no notable increase in carbon in local sediments. And while growing seaweed sounds harmless, on this scale it has its own risks; it could block out sunlight from the sea surface and suck up all the nutrients for plankton, a side effect that could reduce carbon uptake overall. Despite these limp prospects, scientists still think it is worth giving it a shot.

As climate interventions go, there is something rather appealing about approaches that simply enhance the ocean's natural carbon storage potential, whether that involves a heavy sprinkling of iron into the ocean or cultivating seaweed en masse. But the longer-term side effects of ocean fertilisation are as unclear now as when Russ George first set sail for the high seas. If we want a safer way of exploiting the ocean's biology to boost carbon storage, the most natural method (also investigated by NASEM) is the restoration of ecosystems, an idea that we test informally every time we leave nature alone. The logic is that safeguarding marine life – in sanctuaries, for example – and allowing populations to rebound might store as much carbon in the deep sea as artificial fertilisation, though without the risks. From whales to small invertebrates, marine animals collectively hold about 1.5 billion tons of carbon in their bodies. Restoring just eight species of whale to healthy population levels would store an additional 9.6 million tons of carbon as living biomass in the sea, according to one analysis. What's more, whale faeces are a natural ocean fertiliser as whales can stimulate plankton blooms, by defecating as they travel throughout the high seas; zooplankton either feed on these algal blooms, bringing carbon down deep as they migrate to the ocean's depths, or the whales' faeces become 'marine snow', again bringing carbon down deep. One approach being tested in the Indian Ocean by Sir David King, a former UK government scientific advisor, is whether an artificial analogue of whale poo, wrapped in rice husks discarded by industry, might stimulate plankton. It's a hybrid of ocean fertilisation and ecosystem restoration, which King calls 'marine biomass regeneration'.

Whether boosting plankton, seaweed or whale populations, these approaches all rely to some extent on the ocean's existing biological systems. Another category of intriguing possibilities, looked at by NASEM's experts, involves tapping into the ocean's own physical systems for transporting carbon. Since the 1950s, scientists have looked at tampering with the ocean's natural 'upwelling' and 'downwelling' systems – vertical ocean currents that transport heat and nutrients from the surface to depth and back again – to stimulate fisheries, to generate energy, or even to prevent typhoons. Ulf Riebesell, a German oceanographer at the University of Kiel, is now experimenting with whether they could be boosted to enhance fisheries and to bring more carbon to the deep. His scheme uses pipes fitted with pumps to enhance the ocean's vertical highways.

Ordinarily driven by strong winds, these vertical currents naturally drive deep, cold, CO_2-rich water to the surface and take warmer waters to depth. If successful, Riebesell's approach, which is currently being tested in Atlantic waters off Gran

Canaria, could be deployed in the ocean's nutrient-poor subtropical gyres. The equivalent of deserts on land, these gyres are large current systems that are dotted throughout the high seas, in the North and South Atlantic, the North and South Pacific, and the Indian Ocean. The issue with artificial upwelling is that no one knows whether it would decrease or increase the ocean's ability to store CO_2; if CO_2-enriched water pumped up from depth fails to stimulate plankton bloom, rather than boosting fisheries, this CO_2 would re-enter the atmosphere, worsening climate change. One option to counteract this potential outcome is to couple enhanced upwelling with downwelling, as a way of bringing more dissolved organic carbon to the deep sea. This proposal is equally untested, however, and the only certainty is that it would need a lot of pipes, which have to be constructed, a process that comes with its own carbon footprint.

Elsewhere, scientists have taken the idea of enhanced weathering – that being tested by Project Vesta – and gone further still. Ocean chemist Greg Rau is one proponent of research into a more technologically challenging approach known as 'electrogeochemistry'. As a climate-fixing technology, electrogeochemistry is ingenious. It works by pulsating electricity through seawater to generate hydrogen gas – used as an energy source onshore – and converts CO_2 into the mineral bicarbonate, which is alkaline and counters acidification. With just one process, it generates a non-fossil fuel, stores carbon in the deep sea and neutralises ocean acidification.

By one highly speculative estimate, electrogeochemistry could absorb somewhere between 100 billion and 1 trillion tons of CO_2 each year, depending on how much renewable energy is available to fuel the initial process. For comparison, our current emissions are in the region of 44 billion tons of CO_2 per year

– and growing. What's more the hydrogen would, in theory, generate about 80,000 to 800,000 terawatt-hours per year of clean energy, which compares favourably to our global energy usage of about 160,000 terawatt-hours per year. So, the technology could also meet current global energy demand with zero emissions. That may all sound far too good to be true and, of course, there's a catch. Several, in fact.

The process requires considerable amounts of electricity in the first place, which would have to be renewably generated electricity to qualify as a viable method of CO_2 capture and storage. Even then, to achieve a fraction of its potential, electrogeochemistry would likely require tens of thousands of hydrogen-generating power plants to be deployed, either on shorelines, in coastal waters or on mobile platforms on the high seas. Who is going to finance and construct these plants, and at what environmental cost? Since Elon Musk announced his $100 million XPRIZE for carbon removal, a group of fifteen milestone winners – a mix of academic research groups and entrepreneurial start-ups – have been awarded $1 million each. Of these, three are focused on ocean-based CO_2 removal, and one is Greg Rau's start-up Planetary Technologies, which will use the cash prize to further develop electrogeochemistry as a solution to both carbon removal and ocean acidification. Another of the milestone prize winners is Captura, a start-up based at the California Institute of Technology, which also uses electrochemistry to capture ocean carbon. The third oceans-focused winner was a non-profit group called the Climate Foundation, which is researching the possibility of enhancing ocean upwelling to promote the growth of kelp forests in the open ocean, restoring fisheries and capturing CO_2. The XPRIZE grand prize winner will receive $80 million in 2025.

While the ideas being proposed for climate intervention are undeniably clever and sound promising on paper, the details of what is required are stomach-churning: to stave off the worst of climate change, will we need fleets of vessels traversing the high seas, deploying olivine, in addition to colossal seaweed farms, thousands of vertical pipes in each ocean basin and newly constructed hydrogen power plants on shore? Whatever mix of technologies we test and ultimately deploy in the future, it is very clear that we are in uncharted territory.

And so, ocean carbon storage remains a difficult arena for raising funds. While serious scientific bodies like NASEM are carefully weighing the options, tech executives such as Eric Matzner are taking matters into their own hands, backed by billionaire philanthropists. Project Vesta, for instance, is now part-funded by Irish billionaire brothers Patrick and John Collison, who founded Stripe, the online payment processing platform. Oceankind, the non-profit that hosted the Half Moon Bay conference, has an anonymous philanthropist backer. Large corporations and foundations such as Google, Amazon, Stripe and the Gates Foundation – all of whom want to bill themselves as net zero – are emerging as donors.

Globally, tens of billions of dollars will be needed in the next decade to figure out the best way forward. One issue for those wanting to monetise and finance their scheme by selling carbon credits from it, just as Russ George did, is verifying any claims they make about carbon sequestration. Another is environmental liability: who is responsible for the negative impacts – should they occur – in international waters? And what about transboundary effects, where a technology deployed on the high seas has impacts in national waters? Iron fertilisation, for instance, may cause havoc at the test site, but the effects will by no means be isolated; by changing the amount or type of plankton in the water, it could spread harmful blooms or introduce non-native species to other nations' waters.

If all of this – electrifying the ocean, dumping approximately 5 billion tons of rock at sea, deploying tens of thousands of vertical pipes, creating fake whale faeces – sounds far-fetched in the extreme, it's worth remembering why we're considering experimenting with these ideas and evaluating their risks and benefits in the first place. Each of these outlandish-sounding proposals has been born out of sheer necessity. These schemes, extreme as they are, are the parachute we might have to rely on as we're running off the cliff. For without large-scale carbon capture, we are unlikely to avoid dangerous climate change – of 1.5°C (2.7°F) or even 2°C (3.6°F) above pre-industrial temperatures – and even knowing this, our emissions are still rising. That leaves us in the current situation, where some of these schemes will be tested and perhaps even deployed, in some form, in decades to come. Even so, it's worth remembering that this might not be enough to stave off dangerous climate change. Carbon dioxide removal is, after all, a slow burner, yielding results on the time scales of decades. To intervene in the Earth's climate rapidly, we'd need to go even further.

ONE OF KELLY WANSER'S clearest childhood memories is a conversation with her mother about what she'd like to be as a grown-up. 'Pope', replied Wanser. Even today, the tech entrepreneur remembers feeling bruised on being told that wasn't an option for her. Her early ambition, however, may seem fitting to those who now accuse her of wanting to play God. Wanser is the CEO of SilverLining, a non-profit that she founded in 2018 to advocate for research into rapid climate intervention. In short,

Wanser wants us to test and develop radical technologies that would reduce the amount of sunlight hitting Earth, cooling its surface, and alleviating some of the harm caused by global warming. The idea is controversial, the technology speculative and, with few backers, the coffers have, until now, been largely empty. Wanser plans to change all of that.

'We have an increasingly fast-moving problem, and we may need increasingly fast-moving countermeasures', Wanser told me, the first time we spoke. Under ordinary circumstances, I would have wanted to meet Wanser in person, but there was a global pandemic. So we settled on meeting virtually – I sat in my home office and Wanser sipped coffee in quiet downtown Colorado. A mix of Silicon Valley tech exec, science nerd and climate intervention activist, Kelly Wanser has, for over a decade, championed the work of some of the world's leading atmospheric scientists. It was during a series of conversations with climate researchers Ken Caldeira and the late Stephen Schneider, both of Stanford University, that Wanser began her informal training in climate science, technology and policy, subjects that she now converses on with indefatigable confidence. Though Wanser advocates for developing a portfolio of innovations to fix the climate, she's most closely associated with a nascent technology called marine cloud brightening, a term that she coined a decade ago when she co-founded a research programme of the same name at the University of Washington. Before Wanser came on the scene, the technique – developed by a group of atmospheric physicists – was called 'solar albedo modification'. Wanser said that it sounded both wonkish and scary in equal measure.

Like much scientific terminology, 'solar albedo modification' was first mentioned in the peer-reviewed literature. As experts began to use the term, it became fixed, embedded. That is, until

Wanser showed up and suggested something less opaque. For the same reason, Wanser insists on using 'climate intervention' instead of 'geoengineering', which, she says, sounds like a scheme to move mounds of Earth around the planet. In practice, marine cloud brightening involves spraying fine salt particles up at clouds over the sea to increase the number of water droplets within them, making them brighter. As brighter clouds are more reflective, they bounce the sun's rays back to space and away from the ocean, cooling the sea surface and the planet. Though in its infancy, marine cloud brightening could – in theory – relieve regional climate problems, such as the coral bleaching that happens when the ocean overheats; it could also weaken hurricanes before they make landfall, or steady the loss of polar sea ice. Deployed over the high seas, it could have planet-wide effects.

Clouds are odd entities: despite their near-constant presence in our lives, they remain one of the least understood components of Earth's atmosphere. Many of us (at least in childhood) have whiled away innumerable hours staring skyward as these unique arrangements pass overhead. On some level, we appreciate their mutable shapes and sizes, which can morph, in just a matter of moments, from a grimacing face, to a teapot to an elephant's trunk. Clouds are nature's sketchbook, a blank sheet on which we can impose our own creations and fantasies. But clouds are much more than that, even to the uninitiated. The wispy, light cirrus of a spring day can lift our spirits, while grey stratus can make for an oppressive, bleak winter afternoon, dulling our disposition. Some clouds last just seconds or minutes before dissipating, while others can linger for days, creating a more enduring backdrop to life's vagaries. The poet and nature writer Ralph Waldo Emerson once described the constantly changing sky – with clouds drifting in and out to play with the light, shade, volume and colour – as 'the ultimate art gallery above'.

In practical terms, however, a cloud is no more than a mass of super-chilled water droplets hanging, suspended in space. At a height of 0.6 to 6 miles above the Earth's surface, water droplets collect around tiny nuclei. These nuclei can be dust or clay from desert storms or forest fires, sea salt from ocean spray, soot from factories, or even aerosols spewed out by erupting volcanoes. Whatever the source, each of these tiny particles is about one-thousandth of a millimetre in size. Once condensation forms around the particle, it swells to about 0.2 millimetres as a cloud droplet, and once it falls as rain, it is larger again; having collided and merged with other droplets, it is a full 2 millimetres wide. That clouds are condensation vehicles, disguised as fluffy floating objects, has been long recognised. 'A cloud is nothing else but a great heap of snow close clinging together', wrote the philosopher René Descartes in 1637.

What has been less well understood is the vital role the clouds play in regulating Earth's climate; depending on their composition, structure and location, they can act as giant reflectors of sunlight, cooling our planet, or as heat trappers, accelerating warming. Each water droplet within a cloud forms a uniquely reflective surface that throws light off in every direction, making the entire entity opaque. The rationale of marine cloud brightening is that by seeding certain clouds with more nuclei, in the form of sea salt spray, the clouds gather more droplets, sending more light back out to space and allowing less of it to reach the ocean below. It's possible that this would also extend the lifespan of each cloud, making it cool the ocean for longer.

It's important, at this point, to say that not all clouds are created equal, nor are clouds spaced evenly above Earth (even a cursory glance at the sky will reaffirm these facts). So, it matters which clouds are seeded and where. From land, it's probably hard to appreciate that the vast majority of clouds – some 70 to 80 per cent – form over the ocean. Most of these are stratocumulus.

The Latin for 'lumpy layer', stratocumulus clouds, from above, look like thin rumpled blankets. They drape over wide stretches of sea, but most often they hug the western coasts of North and South America, and Africa, sheltering their blue waters from the sun's scorching rays. Among clouds, stratocumulus are the strongest light reflectors and have the biggest effect on climate. On average, they are about 66°F colder than the ocean's surface, reflecting between 20 and 30 per cent more sunlight than the waters below. By comparison, thin, wispy cirrus clouds, which form higher up in atmosphere, have quite the opposite effect: like greenhouse gases, they trap heat.

The concept of seeding stratocumulus clouds throughout the ocean to tackle climate warming was first proposed by British physicist John Latham in a short correspondence to the journal *Nature* in 1990, and subsequently developed into a technique that could be tested in the field. By 2008, Latham had teamed up with Stephen Salter, an engineer at the University of Edinburgh, to describe the practicalities of marine cloud brightening. They envisaged an armada of around 1,500 salt-spraying ships covering a quarter to a half of the world's ocean, traversing the high seas, with the sole purpose of lowering Earth's temperature. Each of these wind-driven 'Flettner' vessels, which are propelled using an unconventional vertical rotor, would sail back and forth perpendicular to the prevailing wind, releasing micron-sized droplets of seawater beneath the ocean's clouds; these Flettner vessels would use turbines – rather like oversized propellers – dragged through the water to generate the electricity needed to turn seawater to mist and loft it up into the atmosphere. The futuristic scheme could theoretically increase light reflection from the atmosphere by 1 or 2 per cent, which would offset a doubling of CO_2, equivalent to 3°C (5.4°F) or more of warming.

Others have suggested a more judicious approach, just targeting stratocumulus clouds off the coasts of North and South America and Africa, with the aim of impacting the climate elsewhere; this could work to stabilise Arctic sea ice, weaken hurricanes in the tropical northern Atlantic – to the east of the Gulf of Mexico – and prevent coral bleaching, especially in the tropics.

So far, no one has tested any of this at scale, and there are valid concerns about how such a scheme might affect the ocean and about how brightening clouds in one region could impact the climate elsewhere. Most of the data on marine cloud brightening comes from computer simulations, and so it's still unclear whether tampering with ocean clouds would cool the water below or produce fewer, larger drops, which would produce a regional warming effect, instead. One possibility that shows up in climate models with salt-seeded clouds is that as global temperatures drop, so does rainfall, in some places more than others. One study, in particular, suggests that the Amazon basin would be hard hit by drought, triggering an ecological and climate disaster.

The possibility of there being clear winners and losers is tough on those, such as Wanser, who see the technology as a means of fixing a mess that's been largely caused by rich nations, when the world's wealthy are the ones who will finance, control, and even possibly profit from the technology and its deployment. 'There's this paternalistic notion that, for the sake of developing nations, we need to stop this motion towards engineering the planet. But they may be quite desperate for solutions', Wanser told me. 'If we have measures to keep the system stable, we need to look at them. It is unconscionable not to', she said. While Wanser has spent the past decade advocating for testing such radical interventions, to see if they work, the research has moved at a glacial pace, bogged down in controversy over who gets to use it, where it's deployed and what the unintended consequences might be. (The same, of course, has been true of climate interventions targeting the atmosphere, which is another global commons managed with a hodgepodge of regulations that have, collectively, proven ineffective in tackling the climate problem.)

Now, however, as the options are narrowing, people are beginning to listen. Proponents of solar intervention favour the fact that it could be deployed globally at relatively low cost, lowering temperatures within weeks or months. Its critics, however, point to the fact that it tackles the symptoms of climate change, and not the cause, which could actually encourage us to continue polluting.

Another deterrent is the fact that 'weather modification' – similar in practice but with a less enduring outcome – has been used in warfare. During the Cold War, both the US and the Soviet Union funded research into cloud seeding, an approach the US successfully used during the Vietnam War to extend the monsoon season and disrupt enemy troops by making their supply routes so wet as to be virtually impassable. Though weather warfare is now outlawed by the UN, it's all too easy to imagine that geoengineering, masquerading as some benevolent form of climate intervention, could be surreptitiously deployed for similarly hostile purposes. For these reasons, advocates of carbon dioxide removal generally try to dissociate themselves from those in the camp of albedo modification, which is seen as a more rapid, quickfire and desperate solution to the climate crisis. In truth, we know little about how either of these approaches might play out.

Unlike Russ George, Wanser never actually wants us to use climate intervention. She hopes we'll never have to. By her own admission, it's a drastic act. At the same time, she's pretty sure that the only thing worse than needing to fix the climate is needing to and not knowing how. What makes her stance credible is that she doesn't plan on profiting financially from this. Unlike ocean seeding, cloud seeding doesn't have a commercial angle. There are no carbon credits to be gained or traded. So why does Wanser, a successful tech entrepreneur with more than five start-ups and twenty patents, care at all?

Wanser, who describes herself as a socially minded risk-taker, began her first start-up in 1999, when she launched a viral email marketing business from PARC, the Palo Alto Research Center. Within eight weeks, she had raised $11 million and recruited a team of twenty-five. For her, climate intervention is a way of preventing human suffering, especially for those – such as the inhabitants of low-lying nations – where the impacts are likely to be felt soonest and hardest. 'If the best that we can do with policy is leaving people unsafe or dead, that is not acceptable', she told me.

Now an independent research centre, PARC was, in its first incarnation, a subsidiary of Xerox established in 1970 and dedicated to advancing computer science. It's credited with facilitating all sorts of groundbreaking inventions including the ethernet, laser printing and the personal computer. These days, when Wanser visits PARC, it is to see the 'old salts', a group of half a dozen physicists, engineers and computer scientists who, twelve years ago, came out of retirement to develop the technology for brightening marine clouds. In effect, this volunteer group are putting into practice Latham and Salter's vision in the 1990s of albedo yachts traversing the high seas, spraying clouds from a series of tall funnels.

For the time being, the 'old salts' are designing a simpler, nozzle-spraying system that can be scaled up as needed. The old salts jokingly describe themselves as 'plumbers', just working out the nuts and bolts of someone else's concept, but it has been a considerable feat of engineering to design even a single nozzle. Each nozzle is designed to deploy 3 trillion sea salt particles a second and each particle needs to be less than 100 nanometres in size. On top of this, they must also contend with the fact that seawater is corrosive, especially at such high volumes.

Over the years, the old salts have tried and tested all sorts of materials out of which to form the nozzles including diamond and sapphire, but the salt just destroyed everything. Eventually, after much trial and error, they settled on an electromagnetic material. In March 2020, researchers conducted the first field trial on actual clouds – using the technology developed by Wanser's programme – over Australia's Great Barrier Reef, in a bid to save the UNESCO World Heritage site from another mass bleaching event triggered by global heating. The trial will now need to be scaled up to draw any meaningful conclusions about its usefulness for climate intervention.

ABOUT A SIX-MINUTE walk from PARC, also in Silicon Valley, is Stanford University. Its sprawling campus has long been the backdrop of salient climate conversations, including those that first set Wanser on her path as a climate intervention activist. It's also where Leslie Ann Field, engineer and innovator, has set up shop. Like Wanser, Field believes that we may, one day, need to cool the ocean to fix the climate. But she has other ideas about how we might achieve it.

Field is the founder of the non-profit Arctic Ice Project, an organisation whose purpose is to investigate how technology could restore thawing sea ice. Without such intervention, in just a few decades from now the Arctic's Fram Strait, a deep, frozen passage that connects the high seas of the Arctic to those of the Atlantic, will be transformed into a global shipping route that will ferry goods between Asia and Europe in record times. Field's vision of its future is very different indeed. Her pitch is that by refreezing Fram Strait's sea ice, she can counter the Arctic's, and the planet's, rapid warming. She views polar sea ice as a safety lever in the climate system. By rebuilding it, she says, we can prevent the runaway climate change that would otherwise be spurred on by the loss of this protective cover. 'The reality is that for hundreds of thousands of years', she told me, 'we've had this heat shield in the Arctic reflecting the twenty-four-hour-a-day summer sun. And now we don't. It's disappearing over large areas in the summer.'

Field's concept is simple. She proposes scattering tiny glass beads on the surface of thin sea ice, as a way of boosting its reflectivity, helping it to regrow. In 2008, after watching the climate documentary *An Inconvenient Truth*, Field had an epiphany that she could devise a method to make thin sea ice more brilliant, more dazzling – like a fresh pile of snow – and importantly, more reflective. When sunlight hits thick ice covered in fresh snow, as much as 80 per cent of it bounces off the ice and is reflected into space; with young, thin ice, only 30 per cent of light is reflected, and a lot more heat is absorbed. When sea ice melts entirely, it is replaced with dark ocean, changing the albedo, or reflectivity, of the entire Arctic region. 'Open ocean reflects almost nothing', said Field. 'You can't do better than snow, really fresh snow.' After more than a decade spent researching the options, in 2019 Field filed a patent for her innovation to use glass silica beads, each just 325 microns thick – about the width of a strand of hair – as a material to brighten Arctic sea ice. It was her fifty-ninth patent. The following year, in 2020, she filed another patent, for what she calls the 'ice delivery system', which is a description of the method she plans to use to deposit the glass beads.

So far, Field has scattered these minute silica spheres over frozen lakes and ponds in Canada and the US, including sites in the Sierra Nevada Mountains, Minnesota, Alaska and Alberta. The results have been encouraging. At the Minnesota site, for example, spreading a few layers of silica over thin ice boosted its reflectivity by 20 per cent, which was enough to delay seasonal melting. Cooling the ice overall, Field claims, will also lead to better ice retention, with less of it being lost to warm water melting from below. A wildflower gardening enthusiast, she has a penchant for spending long days outdoors, rigorously testing her methods, while singing arias. In 2020, one of her colleagues described being on site with her in Alaska: 'As we slogged through the boggy tundra in our thigh-high boots towards the research station, Leslie sang opera. And not just any opera, but a "Queen of the Night" aria from *The Magic Flute*. The lemmings and snowy owls were quite impressed.'

Field still needs to prove that her concept for sea-ice restoration is effective for climate intervention and harmless to marine life – a big risk and, currently, a big unknown. From the computer modelling and site work that she's done so far, Field has inferred that it wouldn't be necessary to carpet the entire Arctic in silica. Instead, she would hope to boost the reflectivity of thin sea ice from 30 per cent to 45 per cent in a small strategic place such as Fram Strait, which measures around 38,000 square miles, or just 1 per cent of the total Arctic. The hope is that refreezing a location such as Fram Strait could help slow the thawing of the Arctic Ocean.

'Realistically you want to have as light a touch on the planet as possible. Our first principle is to do no harm. The whole point is to leave things better', Field explained. The scheme, if successful, would cost about $1 to $5 billion annually to maintain. 'It is a lot of money', she admitted, but one which compares favourably to the trillions of dollars of climate-related damage that have already occurred. It's unclear who would approve, or oversee, such a deliberate planetary intervention on the high seas. Like Wanser, Field's reasoning is that her work focuses on a back-up plan that she hopes we will never need. If we do need it, she takes comfort from the fact that silica is abundant in nature – in the shells of marine creatures, for instance – and that the notion is to restore something that is only now being lost. 'The default stance of conservationists – myself included – is I don't want to intervene. We don't want to play God, good grief. Right?' said Field. 'If we don't do anything, nothing that we are trying to preserve will be left in a few decades. There's just this vast risk of doing nothing.'

Influential organisations, and governments, are beginning to take note of the need to consider, and fund, this type of research. Meanwhile, however, climate intervention research is still largely being championed by tech entrepreneurs and funded by philanthropists in wealthy nations – a risky set-up, given the uncertain impacts, some of which may be regional, affecting those least involved in contributing to the climate problem and in these unorthodox solutions. In 2020, Wanser helped to push forward bipartisan legislation to allow for atmospheric climate intervention research to go ahead in the US, the first bill of its kind. Since then the US National Academy of Sciences has called for a national research programme into solar climate intervention – the umbrella term used to describe any effort to either block sunlight or reflect more of it back to space. It includes approaches such as marine cloud brightening and Field's ice project, efforts that would have been unthinkable even a decade ago.

US National Academy of Sciences' president, Marcia McNutt, explained the decision on Twitter like this: 'Solar geoengineering is a last-ditch Hail Mary, to counter unacceptable climate warming. No team plans to be down seven points in the fourth quarter with ten seconds on the clock and 50 yards to go, but teams practise the Hail Mary just in case.' It's a move that Wanser pushed for, and welcomes. Others though remain cautious. For those seeking to protect the global ocean, the idea of it becoming a test bed for our most outlandish ideas is questionable. And many worry about the lack of governance for ocean-based interventions. 'We don't want a wild west mentality, because it can be disastrous', said ocean chemist Greg Rau. 'We need to reassure people that this can be and will be done in a safe and conscientious way.'

6

A 'NEAR-ARCTIC' STATE

IN JULY 2020, the *Xuelong 2* ('Snow Dragon 2') became the first ever Chinese-built icebreaker to head to the Arctic Ocean. The following year, on a second voyage, it reached the North Pole. Capable of penetrating 5-foot-thick ice, working at temperatures of -22°F and of keeping ninety people at sea for months at a time, *Xuelong 2* was deeply symbolic of China's Arctic interests. Though China lacks an Arctic coastline, and Shanghai is located over 4,000 miles from the North Pole, China has declared itself a 'near-Arctic state'. The term has no formal meaning, but everyone gets the message – China has set its sights on the high north.

Its specific plan is a glorious reimagining of China's imperial past, but with a contemporary twist: enabled by climate change, Beijing will forge a 'Polar Silk Road' through the Arctic Ocean, intended to eventually become the route of choice for the world's tankers and container ships. In place of the spices, silks and precious porcelain that travelled along the old Silk Roads, these vessels will carry commodities such as iron ore, oil and natural gas as well as food, car parts and clothing. As the most northerly branch of China's One Belt, One Road Initiative, the Polar Silk Road will be part of a vast logistics and transport network centred on Beijing and connecting Asia, Europe, the Middle East and Africa.

For China's plan to work though, the Arctic Ocean and in particular the Fram Strait – the deep passage that connects the high seas of the Arctic to those of the Atlantic – will need to be ice-free. And so the purpose of *Xuelong 2*'s July 2020 mission was for China to assess, for itself, the rate of Arctic climate change and the speed at which it can proceed. All going to plan, China's Arctic encroachment will have two steps: the first, imminent, phase is to gain access to Russia's existing open-water Arctic sea route, the Northeast Passage.

This scheme demands close ties between Beijing and the Kremlin and, for logistical and economic reasons, the tentative relationship seems to suit both parties. Russia requires Chinese investment to continue expanding its Arctic infrastructure, such as its liquefied natural gas projects. China has needs of its own, including gaining access to Russia's coastal sea route, to fulfil its greater ambition.

Phase Two frees China of its Russian ties. This step will see China use a new transpolar shipping route, running through Fram Strait, and through the Arctic Ocean close to the North Pole. Expectantly named the 'Central Passage' by China, Beijing's ambition is for this route to compete with, and eventually supplant, existing polar shipping lanes – the Northeast Passage, which is controlled by the Russians, and the Northwest Passage, which is claimed by Canada. Uniquely, the Central Passage will cross the high seas, where China, and other nations, enjoy the freedom of navigation. Here, China will no longer be bound by any other nation's rules; its ships won't have to obtain permits from the Russians or pay fees for passing through the Suez Canal. The Central Passage will have other advantages: it will shave as much as ten days off the journey time between Asia and Europe and its deeper waters will accommodate the world's largest vessels. And at the rate that ice is thinning in Fram Strait – currently faster than predicted by climate models – the Central Passage is likely to be as accessible by mid-century as any existing polar sea route.

For the time being, there are easier shipping routes. The Suez Canal through Egypt has – for 150 years – connected Europe to Asia, and these days funnels around 12 per cent of global trade, averaging about $9.6 billion worth of goods per day. The Central Passage would cut around 2,700 nautical miles off this Europe to Asia journey. In taking the shorter route, ships will use less fuel, but their presence in the Arctic is not without consequence: like the rest of the high seas, these waters are governed by an 'out of sight, out of mind' philosophy where rules are often voluntary, and enforcement is lax.

In recent years, ships flying flags of convenience, with laxer enforcement of environmental and safety rules, have started to encroach upon the Arctic. One analysis found that six of the seventeen foreign ships that transited through the Arctic high seas in 2018 were not suitable for polar transit, posing an oil-spill threat, an incident that would spell disaster in the frigid waters of the high north. What's more, rules that apply to Arctic operations to limit environmental damage, notably the International Maritime Organization's Polar Code, are mostly voluntary and Russia has its own rules for Arctic shipping, some of which have yet to be reconciled with international standards. For example, container and carrier ships often use a heavy-grade oil that releases CO_2 as well as black soot. In Antarctica's Southern Ocean, there's a ban on this fuel, but in the Arctic, no such ban exists. In its absence, it's a near certainty that a growth in Arctic shipping will darken the surrounding ice and snow, accelerating warming further.

The year before *Xuelong 2*'s maiden Arctic voyage, I headed to Fram Strait with Greenpeace to meet a team of polar scientists who were there to measure the changes underway, and to better grasp the future for this remote part of our planet. For the non-profit, it was the first stop on a twelve-month expedition through the high seas, designed to support scientific research focused on this planetary frontier, and to raise public awareness of the need for its protection. By the time of our visit, in 2019, the Arctic region had already warmed by 3.6°F since the pre-industrial era, more than twice the global average. The rate of change was so striking that its marks could be seen everywhere. Just before our arrival, the area covered by sea ice had reached a new record seasonal low, roughly 10 per cent below the 1981–2020 average, topping an already vexing trend. In just four decades – since the start of the satellite era – Arctic summer sea ice had shrunk by 70 per cent in volume.

Nowhere were the changes more pronounced than in Longyearbyen, a small coal-mining town on Svalbard's Spitsbergen island and the departure point for our expedition. Just weeks before our arrival, the town had entered its 100th consecutive month of abnormally high temperatures, which at times were as much as 12.6°F above historic records for the season. On land, the permafrost was melting and was, somewhat ironically, threatening the Global Seed Vault – a repository of seeds important to food security and itself a response to a potential climactic apocalypse, built into the frozen mountainside overlooking the town. Longyearbyen, which had long prided itself on being the world's most northern town – home to the most northern petrol station, hotel and supermarket – had suddenly been rebranded as the world's fastest warming town. Its residents were already experiencing changes that the rest of us are unlikely to face before the year 2200.

Fram Strait lies west of Spitsbergen island and – as its direct neighbour – is also feeling the heat. Temperatures here have shot up by 7.2 °F since the pre-industrial era. Among scientists, there's little doubt that the Strait, and the Arctic Ocean, will eventually be ice-free in summer. The question is when. The last global effort to find out came up with 2050 as a likely date – assuming greenhouse gas emissions remain high. But more recent research has brought the date forward, suggesting that Arctic summer sea ice may even be a distant memory by 2035, around the time my kids are entering their twenties. By one estimate, when this happens, it will mark the first time in 2.6 million years that the Arctic Ocean has lacked any sea-ice cover.

MY STOMACH HEAVED as I prepared to climb aboard the helicopter from the deck of Greenpeace's *Esperanza*, a former Soviet fire-fighting ship that had been repurposed as a research vessel. We were to head north, tracking the edge of the Arctic sea ice to the east of Greenland. Outside, the temperature was 23 °F, warm for a spring day in the Arctic, but still bitterly cold to an unseasoned visitor such as myself. Also boarding the chopper was Knut, an avid birdwatcher and hiker who lives in Finnmark, Norway's northernmost county. Knut was, I suspected, more adept than me at dealing with the weather, a suspicion that was quickly confirmed by his preparations on deck.

'Slå floke', Knut mouthed as he waved his arms back and forth across his chest and jumped up and down in a gesture that, he informed me, was standard protocol for Norwegians in the cold.

Beneath my survival suit and life jacket, I had on multiple layers – thermals of the sort used for skiing, topped with a tracksuit and a thick down jacket. I squeezed my head and hands through the tight rubber rims of my survival suit, then quickly stuffed my half-frozen hands inside a pair of oversized gloves before I, too, joined in the 'Slå floke'. Despite the constraints of our clothing, it did the trick.

Soft snowflakes drifted into the hangar, settling on our boots, suits and exposed faces. I brushed them away, and swiftly pulled on my balaclava, snood, life vest and woolly hat before placing my final bit of kit – a pair of black ski goggles – on my head, waiting for the signal to board. To quash my nerves, Knut and I took turns spotting seabirds idling on the breeze – first a fulmar and then a pair of glaucous gulls who were among the Arctic's early summer visitors. After what seemed like an eternity – during which the crew were running through a crash safety drill – we got the nod. I repositioned my goggles, pulled my snood up over my nose and slipped on my headset and mic.

Once airborne, we'd fly directly over Fram Strait. Our mission was to search for solid slabs of ice – 'floes' – that looked large and stable enough to hold a team of ten – scientists, photographers and crucially, our polar guard, Tim. All we had to do was keep our eyes peeled. But I was nervous. The previous day, while discussing the helicopter mission or 'heli-ops' with one of the expedition's biologists, Hillary Glandon, she confessed that neither she nor Till Wagner, the polar physicist overseeing the entire mission, had any intention of boarding the chopper. Neither fancied their chances, both reminding me that they have young children at home, that the safety stats aren't great for helicopters, and that there's little chance of being rescued at 80 degrees north. Still, I figured everyone on the expedition had the intention of returning to somewhere, so I decided to take my chances and took my place in the front, next to Marty. A seasoned pilot from Northern Ireland, Marty had been flying for twenty-five years, and, for him, this was just a routine mission. As we took flight, Marty sensed my nerves and tried to distract me by admiring the view and making chit-chat over the mic about recent musicals he'd seen back home. The reception was poor, his voice muffled against the din of the chopper's blades and the air heater. I misunderstood and, for one long moment, thinking he'd partaken in these musicals, I admired his versatility.

Soon after taking flight, we found ourselves 79 degrees north and several thousand feet above our two ships, Greenpeace's *Esperanza* and *Arctic Sunrise*, from where I had a bird's eye view of a changing Arctic. Below us, patches of thin 'grease ice' formed what looked like long oily slicks. Occasional solitary icebergs glistened against the inky blue-black ocean, turning iridescent aquamarine where they dipped below the waves. Further north, the ice thickened and formed loose clumps that, from the air, looked like heaps of meringue smothered with cream – one giant Eton Mess. Ahead, in the distance, near latitude 80 degrees north, we spied solid white terrain, scarred here and there by deep blue rivulets. Dotted on the ice itself were snowy bumps and lumps, some the exact size and shape of polar bears, tricking us into looking twice. Directly below us, the mush of meringue suddenly joined to form larger slabs, each over 30 square feet in size. These were the floes we were after. To look down on the Arctic is to view a system in flux, one straining to reach a new equilibrium. I was awed and soothed by the beauty of this spartan, serene seascape, but as an interloper, I couldn't help wonder whether each break in the pack ice, each slick of grease ice, was seasonally typical or something more sinister, indicative of a wider trend. I wasn't the only one concerned. As Hillary Glandon put it: 'If you lose the ice then the whole system falls apart.'

FOR MUCH OF modern history, the Arctic Ocean has been an unnavigable frozen wilderness, considered largely useless from a commercial perspective. That's not to say that there haven't been attempts to forge a way through. The first were in the sixteenth century at a time when the Spanish and Portuguese still claimed much of the ocean as their own, including the main sea routes to Asia. Forced to establish their own passageways but unable to penetrate the thick year-round ice of the Arctic, British and Dutch sailors tried to navigate the icy coasts of present-day Canada and Russia. Exploring these frozen waters, however, came at the cost of many lives. In the mid-sixteenth century, Olivier Brunel, a Flemish navigator, began to explore the possibility of a sea passage to China along the Russian coastline, but was drowned in Russia's Pechora Sea. More than three centuries later, in the summer of 1879, a Finnish-Swedish research team eventually became the first to successfully navigate the Northeast Passage, on a journey that took them two years. Their steamship *Vega*, constructed as a whaling and sealing ship, became trapped in the sea ice about a mile from the Chukchi Sea at Neshkan, Russia, and only ten days from their final destination in the Bering Sea. Forced to overwinter until the ice melted, the voyage took two years to complete.

Around the same time that the British and Dutch were seeking a sea route via Russia, an English privateer named Martin Frobisher convinced a group of merchant traders to fund his exploration of a fabled Northwest Passage, through the Canadian Arctic. Though unsuccessful, his three failed attempts fuelled a greater ambition among explorers to locate the fabled route. Confident of his abilities to fulfil this quest for the English crown, Captain James Cook came out of retirement to attempt the Northwest Passage in 1776, only to meet a gruesome death in Hawaii, before even reaching his starting point in the Bering Sea. Three-quarters of a century later, polar explorer John Franklin also met his demise in his search for the Northwest Passage: his third attempt lost both HMS *Terror* and HMS *Erebus* to the sea ice and, with them, all 129 crew. Eventually, the Norwegian explorer Roald Amundsen claimed this victory in 1906.

Crossing the high seas of the Arctic Ocean, however, was an even more sobering proposition. The first serious attempt came in 1893, when the Norse explorer Fridtjof Nansen began a voyage that took his ship, the *Fram* (or 'Forward' in English), through the narrow channel, the Fram Strait, that still bears its name. Though Nansen returned to Norway a hero, honoured for his bravery and tenacity and for reaching the furthest point north, frustratingly he and his crew ended their mission 260 miles south of their destination. It wasn't until August 1977, the year of my birth, that a Soviet nuclear icebreaker named *Arktika* became the first surface ship to break through the pack ice and make it all the way across the solidly frozen Arctic Ocean to the North Pole. Departing from the industrial city of Murmansk, *Arktika*'s voyage was timed to coincide with the sixtieth anniversary of the Bolshevik Revolution. Within my lifetime – and quite possibly a mere sixty years after *Arktika*'s conquest – this same voyage will become routine, as the Arctic Ocean becomes ice-free.

In a location as remote as Fram Strait, it can be difficult for scientists to collect data often enough to make any meaningful assessments about how, and why, the region is changing. But several hundred miles west, and one degree north of Longyearbyen, was one such opportunity. In 2014, an American oceanographer named Mattias Cape led an expedition that laid a transect – essentially a straight line marked at predefined intervals – along the 79th parallel north, running through Fram Strait all the way to the edge of the ice sheet at Svalbard. The point was to look at the physical characteristics of the water – whether it was layered, mixed, warm, cold – as well as what was living there. What Cape and his team discovered was fascinating, if alarming: their study showed that the Arctic Ocean was, over time, becoming more like the Atlantic Ocean.

Dubbed the 'Atlantification' of the Arctic, the intrusion of warm Atlantic water from the south, in part due to changes in ocean circulation, is literally melting Arctic sea ice from the bottom up, and fast eroding the glaciers that hug Greenland's eastern coast. It's too early to say how quickly this will evolve or even why it is happening. In a climate-altered world, it's tempting to pin such changes on human activity, but for that, scientists need evidence and years of it. In scientific terminology, the sort of evidence needed to show that the Atlantification of the Arctic is an ongoing (never mind a human-caused) trend is called a time-series. Right now, the time-series stretches back to the late 1990s, composed of occasional studies like Cape's, as well as data from tethered moorings maintained and checked by institutions such as the Alfred Wegener Institute in Germany. But the data are not enough. 'This is a region that's undergoing a lot of change, and from different directions. It's complex and it's remote, so we have little information to go by', explained polar physicist Till Wagner. One reason we had come to 79°N was to add another data point to the time-series.

On our outward journey from Longyearbyen, it had taken our two ships, *Esperanza* and *Arctic Sunrise*, ten hours of constant steaming to arrive at the inconspicuous patch of water that was our starting point. I'd been allocated a berth in a cabin shared with three others on '*Espi*' as the *Esperanza* is affectionately known by its crew. *Espi* was stable and renowned for giving sailors a good night's sleep; in comparison, *Arctic Sunrise* was nicknamed 'the washing machine'. Acquired by Greenpeace from Danish sealers, it was a purpose-built icebreaker whose V-shaped hull made for a particularly queasy ride.

During the long outward journey, the crew prepared *Espi*, and ran us through numerous briefings and safety drills, showing us – for instance – some of the 140 fire extinguishers that are a remnant of its Soviet past. Between scheduled orientations and mealtimes, I tried to squeeze in some exercise to offset what would be a fairly sedentary fortnight at sea. At first, I made good use of a punchbag on the deck. Soon, however, the sea turned too choppy to be alone outside, so I was forced to move indoors to the gym. Another reminder of *Espi*'s past, its walls were lined with heavy, rusting weight plates and dumbbells. Before too long, I discovered that rowing indoors in a swell is nausea-inducing, and that weight training in high winds is even more precarious. I finally gave up and decided to do what the crew were doing – relaxing in a Soviet-style sauna near the front of the ship.

It was here that I fell into conversation with Francesco Pisanu, a French deckhand who, in 2013, ended up in a Murmansk prison for being part of the 'Arctic 30', a Greenpeace protest group that the Russians accused of piracy. Francesco goes by the name Frankie, has kind, almond-shaped eyes and a broad grin. I had heard of the Arctic 30, but unaware of the details, I listened intently. Frankie told me that in August 2013, he was working on *Arctic Sunrise* when they left from Norway on a month-long mission to stage a protest against the expansion of Arctic oil drilling. Their plan had been to draw attention to a new platform that had been established by the Russian oil company Rosneft and ExxonMobil, in Siberia's Kara Sea. Forced away by the Russian navy, however, they headed to the Pechora Sea, a stretch of water that runs between the Russian mainland and the archipelago Novaya Zemlya. Once there, two activists tried to board a Russian platform, this one operated by state-owned oil and gas giant Gazprom. Their plan, once at the top, was to reveal their message to the world: 'Save the Arctic'.

As the activists started their vertical ascent of the rig, the Russian coastguard arrived and warned them off, first verbally and then by firing warning shots from AK-47s and from a cannon on their patrol boat. Undeterred, the activists continued their ascent, holding on tightly to the cold metal even as the Russians doused them with water, trying to shift them from the structure. Eventually they were physically removed. The drama seemed to be over, until the following day when the crew of *Arctic Sunrise*, by then in international waters, heard a helicopter circling overhead. Several masked Russian commandos leapt out, seized control of the ship and brought it to Murmansk in Russia. All thirty on board – the activists, the crew, and a journalist – were imprisoned at the SIZO-1 detention centre where they were charged with hooliganism and piracy, a charge that – in Russia – can carry a fifteen-year sentence.

During their two-month stay at SIZO-1, the Arctic 30 were scattered among the general prison populace, beaten by guards and exposed to squalid conditions, including inmates with TB. As Frankie relayed the events to me, I sensed that none of this had been easy. He told me that some of those who'd been arrested still found it difficult to talk about their time in prison. The *Espi*'s radio operator Colin Russell, who is now aged sixty-nine, felt that he could have died in there. As Frankie and the others endured the daily grind of life in a Murmansk prison, news of their plight began to spread, sparking an international outcry.

In Switzerland, football fans protested Gazprom's sponsorship of one of the teams by flying a Greenpeace banner from the stadium roof during a Champions League match; in Germany, activists chained themselves to petrol pumps that were Gazprom branded. In Oslo and elsewhere, crowds gathered, carrying banners that read 'Free the Arctic 30' outside Russian embassies. Eventually, Vladimir Putin caved to international pressure and granted them amnesty alongside the girl-band Pussy Riot, ahead of the Sochi Olympics. Two years later, in 2018, an international arbitration court ruled that Russia had breached the law of the sea by seizing the Greenpeace ship on the high seas and detaining activists, who had protested peacefully, and had the right to freedom of expression. So far from land, in a place beyond the care or jurisdiction of any one nation, it can take years to reach a just and equitable outcome, at which stage irreparable damage has been done. On hearing the story of the Arctic 30, I was struck by how that seemed as true for humans as for the environment being threatened by our continuous need for extraction.

But as Frankie relayed these events to me in the sauna on *Espi*, Russia was already on track for its most successful year ever for oil production – 615 million tons in 2019. Recently, a spokesperson for Russia described its policy in no uncertain terms: 'Plainly speaking: all that can be extracted must be extracted and sold.' As the Arctic sea ice melts, and sea routes open up, this plan becomes easier – drill ships can penetrate deeper into these northern waters, extracting more oil for us to burn in what is quite possibly the perfect example of what scientists call a 'positive feedback' loop; melting enables more oil drilling, which hastens warming.

IF YOU CONSIDER the Arctic Ocean, it's shaped a lot like a bathtub with several drainpipes. Fram Strait is the main exit, or drainpipe, at the southern end – it's the escape route for 80 per cent of the ice exiting the Arctic Ocean into the Atlantic. Except that now, warm water is coming back up this drainpipe, melting the ice from the bottom up. But something unusual, and unexpected, is happening at the other end too: sea ice is being shunted into the pipe at an increasingly alarming speed. And the cause of this has only recently come to light.

Typically, sea ice is born in 'nurseries' in the Arctic's shallow marginal seas, such as the Kara Sea and the Laptev Sea, along the Russian coastline. In winter, temperatures here fall as low as -40 °F and strong offshore winds push new ice out into the Arctic Ocean as soon as it's formed, freeing up space in the nursery. With the right conditions, this ice-producing factory runs like clockwork, sending millions of individual floes fanning out across the central Arctic Ocean, in all manner of shapes and sizes. Pentagons abut hexagons and octagons in a giant geometrical puzzle. Eventually, most of these floes get swept up in a current, the Transpolar Drift, that carries them southwest towards Fram Strait. This same process has happened throughout human history, but now, ice reaches the Strait in a more fragile state than ever before. Only two decades ago, half of the sea ice leaving the nurseries survived the arduous journey across the Arctic Ocean. Now that number has shrunk to 20 per cent, and by the end of the journey, what's left is fragmented, battered and thinner. The rest has melted into the inky darkness, having never fledged properly or having been ambushed en route by warming seas.

Ordinarily, the sea ice born in coastal nurseries would become caught up in the Transpolar Drift, which flows through the Arctic and circulates ice southward out of the Fram Strait. In the place of this nursery ice, sea ice that forms in the Central Arctic Ocean is now hitching a ride on the Transpolar Drift. This conveyor belt of ice is now moving more quickly because the ice is thinner and moves with greater speed. This fills the drainpipe, and quickly – these mobile, light floes toboggan their way across the ocean. For the time being, this means that summer sea ice is bountiful in Fram Strait, even while it's being lost from the Arctic at a rate of more than 10,000 tonnes per second. At first glance, this makes it easy to think of Fram Strait as an Arctic anomaly, a frozen fortress surrounded on all sides by the chaos of climate change: a warm Atlantic intruding from the south, an ocean current, the Transpolar Drift, on overdrive to the north, to the west Greenland's glaciers, creaking at the seams, and to the east, the world's fastest warming town.

But the truth is far less comforting. Till Wagner, the climate physicist and head of all things icy on the mission, cautioned against seeing this as a positive trend. 'This is unlikely to last', he said. 'Ice loss is at an all-time high in the Arctic right now.' Compared with multi-year ice – which has survived several summers and laid down extra layers each winter – thin, first-year ice gets bashed about, broken up, swept along in sea currents and eroded by an intruding Atlantic. The upshot is that in the nineteen years from 1998 to 2017, sea ice in Fram Strait became 30 per cent thinner, altering the delicate balance of this intricate system. One of the theories that Till wanted to test was whether the sea ice being flushed through the Strait was still thinning, and picking up speed. 'If the exit is accelerating, that means we're losing ice more rapidly from the Arctic high seas.'

To find out if this was the case, we needed to get out onto the ice. For me, this meant kitting up for another chopper ride, this time to *Arctic Sunrise*, where I'd get a safety briefing, and swap my immersion suit for 'extreme weather workwear' – something similar to an oversized ski suit, that was large enough to accommodate several underlayers. But just as we were ready to depart the *Esperanza*, the weather picked up, leaving us vessel-bound. With winds gusting at 40 knots and a massive swell lashing at the bow of the ship, I retreated to my bunk. Over on *Arctic Sunrise*, work ground to a halt as the scientists and crew bunkered down. Eventually, after a day and a half of riding out the storm, the winds subsided and we got the all-clear for the following morning.

Over breakfast, however, Fanny, the communications lead for the expedition, kept reminding us that Arctic weather changes incredibly quickly, as if anticipating a problem and trying to soften inevitable disappointment. The logistics of getting to the ice floe to take measurements were complex, and each stage was weather dependent, meaning that our plans could be interrupted at any moment. Sometimes, expeditions can be halted for days or weeks at a time, if winds blow hard enough to keep a crew vessel-bound. First, the winds had to be light enough for the helicopter pilot and the ship's captain to agree to 'green deck' – meaning it was safe to take off from *Espi* and to land on *Arctic Sunrise*. Anything above 25 knots of wind would nix that possibility. From *Arctic Sunrise*, we were to board the Zodiacs to reach the ice floes. But a rough sea would rule out launching the motorised dinghies, because of the difficulty in lining them up against the ship and boarding everyone safely.

Despite Fanny's cautioning, the weather stabilised and held out. Before leaving, our polar guard Tim gave us one last safety briefing. 'Keep an eye out for bears. Just remember that if we see a bear, we leave immediately. Do not try to save your cameras, your ice core drills or your cores. Just run for the nearest boat and hop in', he said. 'But our biggest, most serious threat is the cold', he cautioned, seriously. 'Put on as many layers as possible while still being able to function. Keep your gloves on and keep your balaclava over your face.' He reminded us to keep an eye on each other, looking out for any listless behaviour, which could indicate hypothermia.

On reaching the first ice floe, Till was visibly excited. An erudite German, Till had lost any whiff of accent during ten years of studying maths and physics in the UK, both at Oxford and Cambridge. He now speaks in crisp English tones that he's managed to retain during a decade in the US, first at Scripps Institution of Oceanography in La Jolla, San Diego, and now at the University of North Carolina Wilmington. Over the years he has taken a rather niche interest in how thin sheets bend and stick, and found a practical use for it – to understand exactly how Arctic sea ice is changing.

Out on the ice, the temperature was bearable, at around 5°F. I was double-gloved, but the more experienced among us, including Till, managed with fewer layers and without gloves. As a safety measure, our guard Tim marked a perimeter about three feet from the edge of the floe. Beyond this line was forbidden territory; here the ice could easily dislodge, taking us into the frigid waters below. Tim then piled all of our gear onto a hummock on the ice, from where he positioned himself as armed sentry, on watch for polar bears. Inside the perimeter, Till made ten hollows in the ice, marking the points that he would drill, and others that he would core for later analysis. Till took the first ice core, some 6 feet deep, letting out a happy yelp when it came out smoothly. He then sliced it into 4-inch rings and bagged it, for later thawing onshore. Measured by tape, through a drilled hole in the ice, the depth was around 10 feet deep at the thickest point, tapering off to just under 6 feet nearer the edges. As suspected, it was thin, first-year ice. And it was melting.

THE ARCTIC OCEAN has evolved over millions of years as an ice-dependent ecosystem. Its unique collection of plants and animals are adapted for life in the cold. At the base of the Arctic food web are sea-ice algae, specialist life forms that grow in pockets of water near the bottom of the ice. Adapted to living in low light conditions, sea-ice algae are 'like the shadow plants of a forest. They thrive in the half-light', one expert told me. They have adapted to Arctic life in other ways too. 'Just like the seals and whales, the algae produce lipids – they put on some fat – to protect themselves', she said. 'Everyone has to use some insulation to survive here.' A rich, nutritious food source, these algae are a favourite of zooplankton which, in turn, become food for bowhead whales as well as for the fish and squid, which are staples of narwhals and beluga whales. It is this bountiful nourishment that makes the drainpipe of Fram Strait one of the Arctic Ocean's richest feeding grounds. But as the region warms, and vast expanses of Fram Strait are exposed to sunlight, this will all change.

A whale specialist, Hillary Glandon was the expedition's resident biologist, overseeing all efforts to sight large mammals in the water and on the ice. With a keen eye on the horizon, those on land – the polar bears, seals and walruses – were easy enough to spy. Spotting the whales that visit these waters – bowheads, belugas and narwhals – was a different matter. In the absence of any visible whales, Hillary's next best option was to locate the plankton. 'Whales don't eat the algae, but they do eat the stuff that eats the algae', she explained. While we had spent the first couple of days of the expedition overhead in the chopper, scouring the edge of the sea ice for signs of life, Hillary searched the water samples. In between 'heli-ops', I joined her in a makeshift lab on board the *Arctic Sunrise.* At the back of the lab – up one flight of stairs and through a heavy steel door – was the poop deck, an outdoor area at the rear of the ship and named after the French term for stern, 'la poupe'. The poop deck was where Hillary deployed various nets and profilers to see what life was floating out at sea.

Peering down a microscope, which was secured with heavy bungee cords, I examined the contents of a haul she'd taken using a plankton sampler known as the 'bongo', named for its paired ring nets. Through the heavy lenses, I could see a bunch of zooplankton copepods, some of them literally stuffed with the green algae. Out on the ice, I'd seen these same algae at the bottom of Till's ice cores, tinging the ice yellow-green, like some oversized Limeade Slush Puppie. Zooming in for a closer look, I could see that some of the copepods had visible oil sacs – little yellow lipid parcels they accumulate by feeding on the fatty algae. Hillary, leafing through a heavy taxonomic tome, identified the fat copepods as Arctic species and the leaner ones as recent southern invaders. I peered through the lens again, searching for the most subtle of differences – skinnier limbs and a few more bristles perhaps; all signs of much larger changes afoot.

ABOUT TEN YEARS AGO, Christian Knudsen, a geologist at the Geological Survey of Denmark and Greenland (GEUS), acquired an unusual-looking rock. Coloured orange and roughly the size and shape of a rugby ball, it had been dredged from the Arctic seafloor by one of Knudsen's colleagues. He was intrigued by its outward appearance, but he was more interested in its interior. So, he cut a chunk off. Inside, the rock was layered with fine lines in deep ochres and reds, as though some grand geological feature had been compressed and squeezed into this strange elliptical stone.

For Knudsen, this was good news. The fine lines suggested that the stone had formed over thousands of years in a quiet environment, just like the site it came from. By doing some fairly simple chemical analysis, Knudsen was also able to age it. He discovered that the rock had been sitting in the same spot in the polar basin for 8 million years – since before the last Ice Age. If the rock's appearance had been different – consistent all the way through, like an ordinary beach pebble or a boulder, for instance – it could have been a dropstone, released at random by an iceberg drifting overhead, and useless to Knudsen's investigation.

The lines inside the rock also told Knudsen something else: it was made of sandstone, folded and crumpled during a mountain-forming event over 470 million years ago. The sand grains it contained were older still, closer to 1.6 billion years. The implication was undeniable: this rock was not new seafloor made from some relatively recent underwater eruption; it was continental and its composition was similar to the landmass of Greenland. The rock was part of the Lomonosov Ridge, one of the most mysterious, and contested, mountain ranges on Earth.

Lomonosov isn't visible on a map and it can't be seen from space, but it is more than 1,000 miles long and it reaches over 2 miles above the ocean floor. At one point, it connects to northern Greenland, at another to Russia, and at a third to Ellesmere Island, a Canadian outpost that sits uncomfortably close to Greenland, which is controlled by the Danes.

Knudsen's rock made a compelling case for Danish ownership of Lomonosov, but similar evidence has been presented by the other interested parties, including Russia and Canada who have been fighting over this energy-rich subsea mountain range for over two decades. As far as these nations are concerned, they all own Lomonosov, a convenient assertion given that the rightful owner will have exclusive access to a large slice of the Arctic's untapped seabed resources. By some estimates, the region could hold up to 13 per cent of the world's undiscovered oil and as much as 30 per cent of the world's undiscovered natural gas. And while the Arctic's most accessible oil reserves are on land, 84 per cent of this energy is believed to be offshore.

Meanwhile, other potentially lucrative resources are being discovered beneath the polar seas too, all of which make laying claim to Lomonosov even more attractive a prospect. One 2018 study, for instance, found evidence of almost 550,000 tons of metals, including copper, zinc, gold and silver, on a portion of the Arctic mid-ocean ridge claimed by Norway. In all likelihood, the Arctic – an area rich in specialist microbes, plants and animals that have adapted to life in one of Earth's most extreme environments – is also a site of yet-to-be discovered compounds of use in industry and medicine. Eight nations legitimately claim territory in the Arctic. Only five – Canada, the US, Russia, Norway and the Kingdom of Denmark (which includes Greenland) – have Arctic coastlines and exclusive use of waters extending 200 nautical miles offshore. Three others – Iceland, Finland and Sweden – are deemed to be non-coastal Arctic states. Beyond that, the Arctic Ocean is high seas and the seafloor is part of 'the Area' – a legal term given to the seabed in international waters. For the time being, this area covers around 1.1 million square miles, roughly the size of Argentina. Its resources – minerals, oil and gas reserves and potent compounds for drug discovery – are common heritage, meaning that everyone should benefit from their exploitation.

Russia was the first to assert a claim to the international Arctic seabed, in what was a rather theatrical affair. In August 2007, three Russian explorers broke the thick cover of sea ice at the

North Pole, climbed into a small submersible and descended 2,700 miles into the abyss. There, they planted a titanium, rust-proof Russian flag on the Arctic seafloor, claiming the North Pole – and its potential billions of dollars of oil and gas reserves – as their own. President Vladimir Putin phoned the seamen personally to express his gratitude. The manoeuvre was a blatant challenge to the UN, whose department of seabed sovereignty had previously informed the Kremlin that it needed more scientific evidence to back its claim of ownership.

'In one way, the flag planting was a stunt, it was spectacle. It had no international legal significance', said Klaus Dodds, a professor of geopolitics at Royal Holloway, University of London, who has followed the Arctic situation closely over many years. 'But on the other hand, it brought to the fore a long-standing concern of many nations, which is "What happens if the Arctic Ocean continues to change?" Something that had been imagined one hundred years ago now looked increasingly likely, which is open water in place of a polar desert with a frozen surface.'

While Russia was the first nation physically to assert its Arctic claim, by that point the race was well underway. A year earlier, Norway had laid claim to three patches of the Arctic seafloor, all detailed in a dense geological report submitted to the international Commission on the Limits of the Continental Shelf (CLCS), the same UN body that denied the Russian's assertion of ownership. Since its creation in 1997, the CLCS has received over eighty extended seabed claims, together covering 1.4 million square miles of the seafloor, an area twice the size of Russia. Each claim takes several years to review.

The burden of reviewing these claims is huge, but that's a problem of the UN's own making. When setting nations' Exclusive Economic Zones (EEZS) at 200 nautical miles, the UN also stipulated a caveat: if a nation could prove that its continental shelf – meaning that part of the seafloor that gently slopes from land to sea – extended further out than its EEZ, it could claim an area, usually up to 350 nautical miles from its coastline, as its own. The waters between 200 and 350 nautical miles from land, above the extended continental shelf, remain the high seas, and here rights to fishing and navigation remain unchanged, but the resources of the seabed – mineral, biological, genetic – become national property.

It can take decades to settle extended continental shelf claims, because first, geological expeditions need to be organised to physically collect data and samples – in the case of Lomonosov, rocks the size and shape of rugby balls from over 2 miles beneath an ice-covered ocean. Then the samples need to be analysed, data gathered and collated, before a diplomatic process begins that may, ultimately, call for more data. Much like collecting data to assess the likely harm from seabed mining, or undertaking tests to gauge the potential efficacy of climate intervention, settling seabed claims is a painfully long process. In the Arctic, the case for ownership is especially complex.

In 2014, the Danes put in a claim to a portion of the Arctic seabed covering nearly 350,000 square miles, including Lomonosov. Just as I was leaving the Arctic in May 2019, Canada was standing its ground over an area measuring 67 million square miles, detailed in over 2,000 pages of text. This too, of course, included Lomonosov. In March 2021, Russia reaffirmed its position – but this time supported by more detailed geology, and claiming a larger area still, at least twice the size of France, taking in Lomonosov as well as the Canada Basin, a deep portion of the Arctic just off the east coast of Canada and Alaska. 'Russia has invested a lot of money in the mapping and scientific work', said Klaus

Dodds. 'And Putin has been very clear that the North Pole will be Russia's. But it won't stop with the North Pole. It will keep going.'

The Russian flag-planting incident ruffled feathers, reaffirming the mantra that the Arctic high seas are a new frontier, a *res nullius*, up for grabs. 'That moment set in train a whole series of narratives that the Arctic Ocean was a no-man's-land', said Dodds. The general reaction was encapsulated by former Canadian foreign minister Peter MacKay who said: 'This isn't the fifteenth century. You can't go around the world and just plant flags and say, "We're claiming this territory."' Needless to say, not everyone shares this stance.

The following year, in 2008, the Arctic's five coastal states – Canada, Russia, Denmark, Norway and the US – known collectively as the Arctic 5, signed a declaration of cooperation in the Greenland town of Ilulissat. Ostensibly, it was a commitment from these five nations to protect the Arctic Ocean and peacefully to resolve any future difference over ownership claims. From a conservation perspective, it looked like a major win. Politically, however, the agreement promoted the idea that the Arctic 5 were nations whose seabed claims, once resolved, would make them the rightful owners and guardians of this space.

The move angered several other nations, especially their Arctic neighbours. Since then, there have been other resolutions, most notably a 2018 agreement to ban any commercial fishing in the 'Central Arctic Ocean' – that portion of the high seas closest to the Arctic 5 – for sixteen years. While this agreement, signed more broadly, including by China and the EU, has again been heralded as a conservation victory, it's also an acknowledgement that, in the future, commercial fishing is likely in the far north. Right now, the ice cover makes that impossible. It's telling, said Klaus Dodds, that Norway wanted a thirty-year ban, and

China wanted a four-year ban, but they met somewhere in the middle.

As the Arctic Ocean thaws, boundaries are being redrawn, and claims are being made, for access, for land and for the region's resources. How different it will all be, just decades from now. Occasionally, I imagine that it's 2050 and I have grandchildren in school, learning, for the first time, about the real and imagined boundaries that shape our world. Their first task is to look at political maps, showing national territories and borders, and to memorise the names of countries and of ocean basins. It's conceivable that by 2050, the Central Arctic seabed – all 1.1 million square miles of it – and its precious resources will have been carved up among the Arctic's rim states. Jointly managed by five nations, this will be accepted as their territory by the rest of the world.

A physical map, by 2050, will have measurably less white for ice, more blue for ocean, while a more complex map, or indeed a satellite image, will show one of the world's busiest sea lanes, the Central Passage, running through an ice-free Arctic Ocean. If the ban on commercial fishing in the Central Arctic Ocean becomes permanent, satellite imagery will show no trawlers, no longliners, no purse seiners (though there could well be rule-flouters here too). But offshore oil and gas rigs might be visible, or support vessels for seabed mining operations.

Climate change is redrawing our political and physical maps, not just in the Arctic but elsewhere. The year after my visit to the Arctic, I found myself again on Greenpeace's *Esperanza*, this time in Antarctica's Southern Ocean, where climate change is altering the landscape and turning the high seas into a battleground of a different sort.

7

THE LAST FRONTIER

ALMOST 150 MILES off the coast of Antarctica lies a tiny ice-capped mountain. One of the windiest places on Earth, its shores are lined with rugged, stony beaches and its sheer cliffs tower hundreds of feet over the icy waters below. Elephant Island doesn't look like much. But in recent years, the waters surrounding this desolate crag – famed for their rich supply of krill, and for the larger creatures, such as seals and penguins, that come here to feed – have become highly contested, a political flashpoint in the conflict over whether to protect or exploit the climate-impacted wilderness of the Southern Ocean.

The waters surrounding Elephant Island have long been synonymous with our exploitation of Antarctica's riches, ever since they, and the greater archipelago of the South Shetlands, were put on the map by a British sea captain William Smith in February 1819. Word of Smith's discovery spread fast, and drew sealing captains from far and wide; using these islands as their base, they harvested 250,000 fur seals for their pelts within three years, before moving on to elephant seals. By the end of the nineteenth century, Britain and the US had taken over a million fur seals from

Antarctica and enough elephant seals to produce over 20,000 tons of blubber. On one island, Macquarie, between New Zealand and Antarctica, over 3 million king penguins were hunted and boiled in giant digesters to extract oil for lamps. Once these populations were suitably depleted, whaling began. All seven Antarctic whale species – blue, sei, minke, sperm, humpback, fin and right whales – were hunted by industrial fishers, most to near-extinction; the humpback, in particular, saw a swift demise, its population plummeting to 10 per cent of historic records.

Commercial sealing and whaling are now outlawed in Antarctica, the former since 1972 and the latter since 1984. By the time these protections came in, however, attention had already turned to Antarctica's lucrative stocks of toothfish, and to krill, a tiny brine shrimp that sits at the bottom of the Antarctic food web and feeds everything from squid and fish up to penguins, seals and blue whales. The Soviets first started fishing for krill in the 1970s – as a response to the Cold War and its desire to replace imported grain – and the Japanese soon joined them, before other nations – North Korea, Norway, Chile and China – came on board. At its peak in the 1980s, the krill fishery was intensive – with catches of up to 550,000 tons per year – and unregulated. Before long, with catch rates declining, concerns started to mount about the large-scale harvesting of krill. In 1982, an international body, the Commission for the Conservation of Antarctic Marine Living Resources (CCAMLR), was introduced to protect Antarctic waters, to manage their fisheries and to safeguard the thousands of unique creatures that live here. But in spite of these protections, Antarctic krill populations are now at 20–30 per cent of historic records – in part a response to climate change – and, while catch rates have declined, interest in fishing these waters has not.

For the fishing industry, the Southern Ocean is one of the few remaining frontiers of the high seas. Countries such as China, South Korea and Russia see krill as a lucrative bounty that fuels a billion-dollar health supplement industry and are now investing in new boats that can each harvest more than 1,000 tons per day. A source of omega-3 fatty acids, which the health industry claims are beneficial for heart and brain functioning, krill is one of the last major underdeveloped sources of protein from the ocean. To conservationists, the Southern Ocean is Earth's last true remaining wilderness, and krill the foundation of its diverse marine life.

What's more, in recent years these have become the fastest warming waters on Earth, and, as such, are in need of greater protection than ever. Around the Antarctic Peninsula – the northernmost part of the continent, which stretches like a crooked finger towards South America – the ocean has warmed 1.8°F since 1955, causing sea ice to recede, and forcing krill to shift southward and to bunch up along icy coasts, where they become easy prey for hungry predators and for wily humans. The conflicts are evident, even in the language used to describe these areas. For conservation planning purposes, the waters to the west of the Antarctic Peninsula, for instance, are in 'Domain 1' – shorthand for a proposed marine protected area that would safeguard all life here, including Adélie and chinstrap penguins, crabeater seals and krill. In fisheries charts, designed by the Food and Agriculture Organization (FAO), this is Area 48 – ground zero for the Antarctic krill fishery.

In the midst of this, Antarctica's wildlife is feeling the squeeze, both from fishing and from rising temperatures, and no species more so than the chinstrap penguin, the region's most abundant seabird. As consummate krill feeders, chinstraps are a bellwether for the Southern Ocean; their status seen as a measure of how this part of the high seas is faring overall. In January 2020, I travelled to the Western Antarctic Peninsula with a group of biologists who were counting these prominent, if little-known seabirds, on remote islands that had rarely, if ever, been surveyed.

By the time of our visit, in 2020, the scientists had reason to believe that chinstrap populations were in steep decline throughout their range. Studies from neighbouring outposts such as Signy Island and Deception Bay had shown a 50 per cent decrease in the chinstrap population over the past half century, a sign that this part of the high seas is changing rapidly. The January 2020 expedition was made possible by Greenpeace, who lent the scientists their boats, equipment and staff. For the non-profit, it was an opportunity to convince people, and especially decision-makers, that the Southern Ocean needs protecting, and that despite its reputation as the world's most pristine sea, its fate still hangs in the balance.

'THE NUMBER OF people who have ever visited this place could fit into a bus, and the number of people who have slept here – well, that's fewer still', said Noah Strycker, as we settled into our camp, a neat row of tents lined up along a narrow beach, next to an unofficial whale cemetery. Pale and wiry, with fine blond hair that has a feather-like quality, Strycker was one of four penguin biologists with whom I had travelled to the Western Antarctic Peninsula in early 2020. Their goal was to survey chinstrap penguins across their known range. Their first port of call was Elephant Island, a former chinstrap stronghold that, when last surveyed by the British in 1971, had at least 100,000 nests.

It was Strycker's ninth Antarctic field season. Since his first visit to the continent monitoring a colony of 300,000 Adélie penguins at Cape Crozier, near the Ross Sea, in 2014, he'd ventured back on numerous occasions as a tour guide, but this time he was collecting data on chinstraps, an accomplishment that would earn him a master's degree, with Heather Lynch, a pioneering penguin biologist at Stony Brook University in New York.

Arguably the world's most committed birdwatcher, Strycker's love of birds began in fifth grade when his teacher suction-cupped a feeder to the class window. Already an obsessive counter of 'things', he soon directed his attention to the winged visitors outside. For the ten-year-old, it was the start of a lifelong passion for which his hometown of Eugene, Oregon, was the ideal setting. Surrounded by coast, mountains and desert, there was a rich variety of birds on offer to watch and to count, and a well-established birding community. 'A lot of birders took me under their wings', he told me, drily. Age eighteen, Strycker was named Young Birder of the Year, an honour given to talented amateurs by the American Birding Association (ABA). Before entering the contest, ABA asks you to first consider the following questions: *Do you never leave the house without your binoculars? Even indoors, are birds always on your mind? If you've answered yes to the above, and you are age ten to eighteen, then you must apply!* Strycker was a clear fit. In 2015, Strycker set himself an ambitious goal: to become the first to see half the planet's 10,000 bird species in a single year. Twelve months and 6,042 species later, he had earned himself a solid fan-base and a Guiness World Record.

During our time together, I found there was no bird-related question I could pose to Strycker that he couldn't answer, a skill that proved very useful for whiling away the hours both on land and at sea. Our journey started in Ushuaia, a small town at the southernmost tip of Argentina, and a common departure point for Antarctica. I had been, rather gratefully, assigned to Greenpeace's *Esperanza*, a sturdy flat-hulled ship with which I had some familiarity. Sailing alongside us was *Arctic Sunrise*, a Greenpeace icebreaker that has a notoriously jovial atmosphere, but was stomach-churning in rough seas. My joy at being given berth on '*Espi*' was not trivial: to reach Elephant Island, we'd first endure a three-day voyage through Drake Passage, the roughest stretch of sea on the planet. The first hazard here is the wind; couched by the Andes Mountains to the north and Antarctica to the south, strong gusts get funnelled through this channel and, with no landmasses to soften their blow, they can be furious enough to send a ship off course. The currents here are equally powerful, and waves regularly reach heights of 40 feet. Even in calm weather, the unbounded nature of Drake Passage gives rise to huge swelling seas, sending ships in a swirling motion that can nauseate the most seasoned of seafarers.

Our departure, however, boded well for the trip ahead. We began our voyage on a calm day in the breathtaking Beagle Channel, named after the ship that took Charles Darwin on his decisive tour of the natural world. From the ship's bow, we spotted pods of Risso's dolphins and Magellanic penguins. As a first-time visitor, I couldn't tell one penguin from another and so Strycker handled the identification, which for him was second nature. After several hours of sailing, we passed Cape Horn, the last piece of reassuringly solid land that any Antarctic visitor sees before they enter Drake Passage. From the starboard bow, I glimpsed its outline, only vaguely discernible behind a blanket of thick fog as we made our way into the gloom.

Rounding Cape Horn is a feat of historical, more than present-day, significance: the world's most southerly sea route, this was how sailors and sea merchants reached the Pacific from the Atlantic before the Panama Canal was completed in 1914. Up until then, the alternative was to navigate the Straits of Magellan, a narrow, exposed and dangerous channel just north of Tierra del Fuego, the archipelago that forms the tip of South America. Drake Passage became the route of choice, but it presented its own hazards: before the advent of GPS, sailors approaching these waters from the west used to see a 'False Cape Horn' tricking them into turning east too soon and heading straight for the Wollaston Islands, just north of the Cape itself. Throughout history, more than 20,000 sailors have lost their lives in Drake Passage. And though crossing the Drake is safe today, it can still be a hair-raising affair that many, myself included, dread long in advance.

All things considered we were reasonably lucky. The crew told me that the forecast for the days ahead was more 'Drake the Lake' than 'Drake the Shake'. The weather was cool and foggy and the seas big and rolling, but everything on the ship stayed in place. In between drills, meals and work, I found precious moments to sneak out on deck or onto the bridge where I could spot birds and whales, usually accompanied by Strycker who patiently taught me to recognise various species of albatross and petrel. Any remaining free time was spent chatting with my cabin mates, including Usnea, an Irish American deckhand, who regaled us with stories of her environmental activist past, a colourful history that resulted in her changing her name by deed poll.

Aged thirty-four, Usnea had, at that point, worked as a volunteer with Greenpeace for several years alongside her twin brother, Silas, often shipping out together and sharing a cabin at sea. Theirs was an endearing and fortunate comradery: spending long periods at sea can be lonely, even if the work is vocational. Most of the crew work in three-month shifts and for many of them, the *Espi* or *Arctic Sunrise* is home and their fellow crewmates family. Life at sea can also be tedious, especially for the crew: days are built around rules and schedules that are strictly adhered to, both for reasons of safety and, I imagine, sanity. Mealtimes take on a significance that I've only otherwise experienced during the Covid pandemic; they are a welcome deviation in an otherwise predictable day. As Samuel Johnson once remarked: 'Being in a ship is being in a jail, with the added chance of being drowned.'

In Antarctica, the need for drills and rules is especially strict, in part because of the punishing climate. The Southern Ocean has some of the deepest, coldest and choppiest waters on Earth; falling in would be an unmitigated disaster, with cold shock and hypothermia taking hold within minutes. Even on deck, air temperatures can be low enough to threaten frostbite or disorientation. Added to this, Antarctica has a unique, and largely unspoilt, biodiversity; efforts to preserve this require all visitors to follow a set of protocols detailed on their permit. I was included on a permit acquired by the scientists and Greenpeace which, from what I could gather, had taken months to obtain. No selfies with the penguins, I had to tell my kids, before leaving. And no unwanted pests. In practical terms, that meant boil-washing my clothes before leaving home. On site, before disembarking the boat we had to remove any visible seeds or dirt by hand and then brush our outer clothes and shoes with detergent. On returning, we'd repeat the routine, scraping away any small residues of dirt and scrubbing down our outer layers until they were glistening. This helped us avoid transferring anything – soil, seeds, bits of grass – between islands. As this was pre-pandemic, none of us was used to the constant need for sanitising ourselves and our belongings; it was all such a novelty that some of us spent hours videoing our biosecurity drills and posting them online.

After three days of watching birds, playing cards and sanitising ourselves, boredom was threatening to set in when Elephant

Island emerged like a mirage, rising out of the misty Southern Ocean. Ahead we could just decipher its stony beaches, craggy cliffs and ice-capped peaks. An inhospitable speck, barely noticeable on a map, this would be our home for the next few days. As excited as I was at the prospect of sleeping on Elephant Island, if I'm being honest, my starting point for the expedition – which I carefully concealed from my travel companions – was that chinstraps are second-rate penguins, compared to emperors, who get all the limelight. I'd been brainwashed – by Disney, in all likelihood – into thinking that, in the penguin world, bigger is better. Around the height of a one-year-old child, chinstraps – which are named after the thin black line that runs under their chin – are rather diminutive penguins. And for me, that meant they were somewhat underwhelming. Strycker, though, had a different view. 'The emperor penguin is gorgeous, amazing, legendary but it wouldn't be the penguin I'd choose to talk to in a bar', he said. 'They are so low-key. Their whole lives are based on endurance, so they are focused on conserving energy. By comparison', he explained, 'the chinstraps are full of life. While the emperors are just standing there, these little guys are running around in circles trying to untie your shoelaces. They are so curious.'

Among Antarctica's penguins – of which there are eighteen species – chinstraps are part of a group known as the brush-tails, which includes gentoos – easily identified by their carrot-orange bill – and Adélies – the classic tuxedoed penguin of the cartoon world. Like their brush-tail relatives, chinstraps have a love of cold water, and of 'porpoising' – a way of travelling that involves fast underwater swimming spurts interrupted every 100 to 150 feet by a quick half-second flight through the air. While most penguins reserve this tactic for when they sense danger, chinstraps seem to do it for fun. But, for three months each year, these agile swimmers are largely land-based. During this time, from about December onwards, they build stone nests on steep cliffs overlooking the Southern Ocean, a strategy that forces them to navigate near-vertical slopes as they make their way back and forth to the sea to forage for krill, their preferred food. This odd arrangement has clearly had its evolutionary advantages – historically, there have been an estimated 7.5 million breeding pairs of chinstraps, a figure that gives these affable seabirds elevated significance in Antarctica where 90 per cent of seabird life, by weight, is penguin. But as the Southern Ocean has started to warm, causing the sea ice to melt and krill to shift southward, the life of the chinstrap is starting to look more like an evolutionary cul-de-sac. While some of their closest relatives – the gentoo penguins – enjoy a varied diet, allowing them to switch to fish and squid when krill are scarce, chinstraps have, it seems, remained stalwart krill feeders, a dependency that is now posing problems as krill populations wane in the waters where they forage.

The day following our arrival at Elephant Island we approached the first rookery. To reach it, we had to climb a steep cliff to a ledge 250 feet high. From there, we had a bird's eye view of one of the island's largest chinstrap colonies. Thousands of penguins gathered in an amphitheatre-shaped space, watching over their fluffball chicks, two per parent, while others splashed in an acrid pool below, stirring up its black, muddy bottom. From where I stood, there appeared to be thousands of penguins, all in a similar arrangement: one parent with two oversized chicks who, by that stage, were about two weeks old. It all seemed so uniform. And, in some ways, it was.

Chinstraps have a co-parenting arrangement in which the male builds the nest, and then the parents take equal turns to fish for their young and to mind the chicks. When one returns, the other departs and so on, until the chicks, at several weeks old, are big enough to enter a 'creche' and the parents hunt together for food. I wondered whether chinstraps – like albatrosses – are strictly monogamous. 'It's more a case of shared real estate', said Steve Forrest, the veteran ecologist of the bunch. 'It's like they've bought a house together and they come back to that same place year after year so, in all likelihood, they end up with the same mate.'

After taking in the vista, the scientists started the count. Strycker climbed to the top of the cliff face, and began counting from the right, using a simple mechanical counter to tally up the numbers in each huddle of penguins. Each click of the counter represented a nest, not an individual, a better measure of the breeding population. Strycker counted each section three times – in small batches – making sure his estimates were within 5 per cent of each other. Alex, a field scientist who was in Antarctica for his sixth field season, began the same work from the left.

Before visiting Antarctica, I'd imagined penguin counting as a fun, easy job, a bit like trainspotting but in a colder climate. I soon realised there was a certain level of fitness involved, and a lot of concentration. Forrest described it like this: 'It's the most focused, exhausting and exhilarating thing you can do outdoors. In any moment, it's like trying to balance a ball on your nose, in that everything else … the wind, the snow, the racket, the state of the world … drops away, and there is only that one thing that matters: nest or not?' As the counting got underway, I tracked a steady line of penguins waddling up and down the cliff face along a highway. Trekking in groups – twos and threes, a cavalcade – the penguins descending the cliff were streaked pink with guano, their dishevelled appearance reminiscent of teenagers at a music festival. The penguins on the uphill climb were clean, clearly having had a dip in the ocean. I was stunned at their mountain-climbing abilities, not to mention their willingness – or ability, at least – to hang out at the top of cliffs, often hundreds of feet tall, in the windiest place on Earth. 'On the windward side it's amazing to see the abuse they can take', said Forrest. 'They can handle a real pummelling.' When they tire of the freezing winds, they plunge into the ocean's icy depths, swimming tens of miles and diving over a hundred feet in search of the extra-salty shrimp that is their choice food.

As the afternoon of the first count approached, Strycker and the team wrapped up their work for the day. Notebooks filled with numbers, they returned to the camp kitchen at Chez Elephant – the affectionate name we'd given our lodgings – where we gathered for coffee and treats. During our stay on Elephant Island, I was regularly reminded of the almost subhuman conditions bestowed upon Ernest Shackleton and his band of men who were famously marooned here in 1916.

One of the first explorers to brave these most southerly of the high seas, Shackleton set off on his third Antarctic adventure, the 'Imperial Trans-Antarctic Expedition', in 1914 with a mission to cross the southernmost continent by foot. But six weeks in, his ship *Endurance* became trapped in the pack ice of the Weddell Sea. Eventually, after several months, the ship was crushed by gargantuan ice floes and so too were the men's hopes of crossing the icy continent. In the end, the twenty-eight men who had travelled with Shackleton had to fight for their survival against Antarctica's unyielding terrain and against the perilous, frigid Southern Ocean. Despite the bleak reality of their situation – which lasted almost a year, including four and a half months on Elephant Island – they survived.

Snacking on some freeze-dried chocolate biscuits in Chez Elephant, I reflected on the stark contrast of our respective experiences, separated in time by a mere century. My situation was undeniably luxurious: at night, we had mattresses to shield us from the hard-pebbled ground and we had a latrine, sheltered within a tent, and supplied with paper. We had cotton liners for our 15-sub-zero-grade sleeping bags and – for those with chilly feet – the offer of a hot water bottle. Even our food was good and plentiful, and there was a constant brew of coffee on the go. In contrast, Shackleton's men slept in soggy sleeping bags placed directly on hard ice or melting snow, and often in garments soaked through from the constant assault of snow and rain. For food, they had some staple supplies, mostly dog pemmican – a type of cold paste of minced meat mixed with fat usually reserved for their animals (which they'd been forced to slaughter and eat). Portions were so tightly rationed, however, that any opportunity to supplement their fare with fresh seal and penguin was a cause of great celebration. On losing *Endurance* to the ice, Shackleton and his men also lost many treasured possessions and more practical items, including toilet paper. They learned to substitute the paper with snow, leaving them chafed, though their excessively meaty diet left them with near-constant constipation. It's unclear which of the two was worse. Whatever discomfort befell us in Chez Elephant – peeing in the open air (the sheltered latrine was reserved for more serious business), being invaded by elephant seals, our clothes smelling of guano – I knew it was nothing compared to what these polar explorers had endured.

As we huddled in the warmth of the kitchen tent, the temperature outside began to drop, forcing us to stay inside and sip coffee while we waited for a sumptuous meal of rehydrated soy protein. The sea remained surprisingly calm for the screaming sixties – a part of the ocean known for its violent storms and extreme waves – and we felt safe on our narrow beach. Eager to know the initial counts, I hassled Strycker for the numbers. It would be a day or so before the biologists had a complete tally of the day's count, I was told, and at least several days before they'd surveyed the whole island. So far, it had been tough. 'A good pace of counting is 1,000 penguins an hour, but if they are dense, it's slower going', said Strycker, joking that he had spent the day counting 'one big penguin blob'. His first section had 1,566 penguins in it, just a tiny fraction of the number on the island. 'I think I'll be seeing penguins in my sleep tonight.'

A VAST FROZEN wilderness, composed of two massive ice sheets, Antarctica is the coldest, windiest and least inhabited place on Earth, with temperatures regularly plummeting to -76°F in winter. Its hostile vista has not exactly welcomed us and it's only in the last 100 years – a mere blip in human history – that we've come to know this forbidding continent at all. But even in that short time, our relationship with this end-of-the-Earth wilderness has evolved into something quite special. Since 1959, ever since the twelve nations active in Antarctic research signed a treaty, Antarctica has been devoted to peace and to science.

It's not only war that is forbidden here but also the dumping of radioactive or hazardous waste as well as commercial industries such as mineral mining and oil and gas exploration. When nations first signed the Antarctic Treaty, they focused solely on protecting land, omitting the surrounding waters. But in 1980, concerned about the rapid expansion of krill fishing, the Antarctic Treaty closed the gap, and created CCAMLR, a body that came into force two years later. With a remit of protecting the Southern Ocean, CCAMLR also regulates fishing, the only industrial activity in these waters. These days, CCAMLR includes twenty-six member nations and the EU, and another ten nations have acceded to the Convention. It's a far bigger organisation than it once was, but it is one in which the balance of power has shifted. The majority of member states are now fishing nations. Around twenty nations fish here, for species such as Antarctic and Patagonian toothfish and for krill. Typically, just over a dozen vessels target krill, in a quota system managed by CCAMLR and divided among seven sub-areas, which cumulatively measure around 1.3 million square miles (roughly the same size as India).

The Antarctic krill stock is managed as an 'Olympic-style fishery', meaning that the regulator, CCAMLR, sets an overall limit for each area and on the total catch, but does not set an individual quota for each vessel. This urges fishers – who, using the most efficient technologies, can suck up over 1,000 tons in a day – to catch as much as possible within a short window, until the total catch has been landed and the fishery closed. As well as each area having its own catch limit, there's an overall cap, called the trigger, which is set at 680,000 tons, a level that is deemed 'precautionary' – in other words, it's low enough to maintain the krill stock for fishers and to support krill predators, such as penguins and whales, in the long term. In reality, fishers don't actually reach the catch limit in a given year, a reassuring sign. But their catch is increasing. In 2020, the year I visited Antarctica, the catch was 495,000 tons, a 15 per cent increase on the previous year.

One issue that concerns conservationists and scientists is that krill harvesting is increasingly focused on small areas close to penguin breeding sites such as Elephant Island. In the 1990s and early twenty-first century, the krill fishery near Elephant Island and the South Shetland Islands started to expand and became a fishing hotspot. By 2014, 90 per cent of the seasonal krill fishing came from this area, and these days the catches around Elephant Island are now more concentrated in space and time than ever before.

Although CCAMLR places strict overall limits on the annual harvest, it hasn't, so far, addressed the fact that driven by climate change, krill are now concentrating in these smaller patches where their predators are forced to compete directly with humans. Naturally enough, as krill have moved, their predators – penguins, humpback whales (which have rebounded from near-extinction to 27,000, close to pre-whaling levels) and humans – have followed them. Take chinstraps for example – they can travel up to 50 miles offshore to forage, but typically they focus on areas where krill swarm. 'The amount of krill taken by the fishery is small in relation to the consumption by penguins and by whales. So, the fishery will not affect the krill stock at the current level of harvesting, but where it takes krill from will be absolutely key', said Phil Trathan, an ecologist at the British Antarctic Survey and advisor to CCAMLR. 'If fishers harvest intensely offshore of a particular penguin colony, then that could affect that penguin colony.'

In the most productive areas, such as around Elephant Island, the catch limit is now being reached year on year, threatening to meet the level reached in the 1990s, before any limits were in place. In 2020, the year of my visit, the catch limit around Elephant Island was reached in 69 days, compared to an average of 130 days over the previous five years. With market demand for krill oil expected to grow, fishers would catch more, especially around those hotspots, if the regulators allowed. These high catch rates pose a particular problem in years when the winter is warm and the sea ice low. A 2021 study, from scientists at the Chilean Antarctic Institute, found that heavy krill fishing during a warm winter with low sea ice tends to result in a smaller population of both chinstraps and gentoos the following year.

Scientists say these are clear signs that the krill fishery is affecting Antarctica's wildlife – a problem for CCAMLR which has a duty both to manage fisheries, and more broadly to protect the Southern Ocean. While there are no reliable estimates of how the krill stock will respond to climate change, one analysis published in the journal *Nature* in 2018 suggests that if greenhouse gas emissions continue to grow and krill fishing expands in the region – a feasible scenario – by 2070 the Southern Ocean ecosystem could look very different indeed. Gelatinous salps (small, barrel-shaped jelly-like creatures) could replace much of the krill, and crabs could overtake toothfish. For their part, chinstrap penguins happily exist in a Goldilocks zone in maritime Antarctica. Further north, they run out of krill and further south, it gets too icy for them to build their stone nests. Their loyalty to this way of life is now proving problematic, as their foraging grounds have become a battleground.

ON LEAVING ELEPHANT ISLAND, I left Strycker and the team to finish their survey while I moved ship to *Arctic Sunrise*, and on to another location, Penguin Island. On *Arctic Sunrise*, I would become acquainted with a new cast of characters, including French chef Laurence Nicoud, who dreamed of rewriting early human history with more dominant female leads, and a German activist named Carola Rackete, who had made headlines the previous July for rescuing a group of Libyan migrants from the Mediterranean Sea. At the time of her arrest, Rackete was captaining the charity ship *Sea-Watch 3*, which she used to bring a number of asylum seekers to a blocked port on the Italian island of Lampedusa. Asked to turn back, she refused, a decision for which she faced a possible fine of $56,000 and a prison sentence. I learned of the story one afternoon when Rackete barged into the campaign room announcing jubilantly that the Italians would not be sending her to jail. Rackete would be our skipper for the journey ahead to Penguin Island, one of the smallest of the South Shetlands.

Once the site of substantial chinstrap colonies, nearing 10,000 occupied nests, Penguin Island was until recently listed as an 'important bird area'. The last survey, completed in 2016, found just over 3,500 chinstrap nests. I was keen to visit, but to reach the island, we needed to brave a treacherously swelling sea. It was the first time in our voyage that the Southern Ocean lived up to its vengeful image. As we boarded the Zodiac – the small RHIB used for travelling short distances – from *Arctic Sunrise*, angry waves threatened to tear each of us, in turn, from the ladder into the sub-zero waters below. As I wobbled on the descent, Rackete pulled me swiftly into the Zodiac and congratulated us all for embarking unharmed. 'Not a beginner's entry' was how she described it. A thick freezing fog descended on our small, rigid inflatable, restricting our view to a few feet ahead. We clung on desperately throughout the thirty-minute ride, faces turned downwards and away from the freezing hail. Finally, Rackete grasped the bowline and heaved the heavy boat, in several short tugs, to a standstill onto the stony shore. Ahead of us lay a shadowy, lunar landscape. A dormant volcano, Penguin Island is defined by its summit, Deacon Peak, which, on that day, cast a red-brown silhouette against a celeste southern sky. As the fog cleared, we made our way along the beach to the first rookery, past groups of resting Weddell seals, moulting elephant seals and fighting fur seals. Approaching the hilltop site, we could see several hundred penguins huddled among the rocks. Much of the site was pebbled ground that would have once held nests, but now lay bare.

Surveying the scene before me, it seemed desolate. Later that day, on returning to *Arctic Sunrise*, I was greeted with news of the first counts from Elephant Island, numbers that reaffirmed the scientists' suspicion: over the past fifty years, some of the largest chinstrap colonies on Elephant Island had declined by as much as 70 per cent, mirroring what had been found elsewhere. 'I think the term "collapse" becomes a pretty reasonable description at this point, you know, when you're looking at more than half of the population having disappeared in such a short period', said Heather Lynch, who was overseeing the research from her lab in Stony Brook, Long Island. She likened the situation to the demise of the passenger pigeon, a bird that was once the most abundant in North America. Known in the mid-1800s for flocks that took hours to pass overhead, darkening the skies and rendering normal conversation inaudible, by 1914 passenger pigeons had been hunted to extinction. In September of that year, the last-known specimen, a female named Martha, died in Cincinnati Zoo. Lynch thinks that chinstraps have remained off the radar until now, because of their abundance. 'We think of there being safety in numbers and there is, to a point', she said, 'but there have been spectacular collapses of species that we thought would never disappear.'

I wondered what lay in store for this part of the Southern Ocean and these islands, which, all things considered, are still largely unspoilt. In principle, CCAMLR has agreed to protect these waters by creating a network of marine sanctuaries in the Southern Ocean. In 2011, its members agreed on the nine areas that would form this network and, in October 2016, the first of these – the Ross Sea marine protected area, a sanctuary covering 600,000 square miles – came to fruition. The world's first properly sizeable high-seas marine protected area (MPA), it protects one-quarter of the world's emperor penguins, 30 per cent of the world's Adélie penguins, around half of the Ross Sea's orcas as well as rare species such as unique sponges that can live for 500 years. One of the most pristine places on Earth, 70 per cent of the Ross Sea MPA excludes commercial fishing.

In 2018, Argentina and Chile joined forces to present a plan to CCAMLR for a marine sanctuary covering the Western Antarctic Peninsula, including the South Shetlands – both Penguin Island and Elephant Island were included in the proposal. Aimed at limiting krill fishing in prime penguin foraging areas, the sanctuary was designed to protect the spawning and nursery habitats of krill and allow the stocks to rebound. But a couple of the major krill-fishing nations, including China and Norway, objected. One of the strengths, and frustrations, of CCAMLR is that all decisions are reached by consensus, so any state can block a motion that doesn't best serve its own interests. As an example, in 2011 South Korea prevented the blacklisting of one of its own vessels that was found fishing illegally in the Southern Ocean. Argentina and Chile resubmitted their MPA proposal in 2020, with Elephant Island excluded. It still hasn't been accepted.

The tug-of-war between the desire to protect the Southern Ocean and the will of certain nations to maintain fishing rights plays out every year in October when CCAMLR meets to discuss proposals for marine sanctuaries. Just three months before I visited Antarctica, proposals to protect the Western Antarctic Peninsula, East Antarctica and the Weddell Sea – which together would cover 1.4 million square miles – were rejected. None of these proposals were new; in fact the East Antarctic sanctuary was being discussed for the eighth year running. Meanwhile, the Southern Ocean is changing more rapidly than any other stretch of sea on our planet, putting its inhabitants – such as the chinstrap penguins – in a perilous position. On journeying to Elephant Island – a tiny crag no bigger than fifty football stadiums – I realised that its fate was part of a much bigger story, of victories that have been won and of battles that are still being fought over the high seas. At times during my stay in Antarctica, the most pristine place I'd ever encountered, I reflected dolefully on how our need for extraction never ceases; it only grows and shifts in focus and location; we stop whaling, only to start krill fishing or seabed mining. It's a certainty that can dull one's optimism for a more sustainable future.

The day after our visit to Penguin Island, I headed out on the Zodiac again, this time to watch whales off the coast of Greenwich Island, another snow-speckled haven among the South Shetlands. It was 18 °F, the ocean a steely pigeon blue. Ahead of us, sheer vertical glaciers lined the coast; occasionally a chunk would break off sending a thunderous roar towards our tiny, insignificant craft. Some snowy sheathbills and a cape petrel flew overhead. Mini bergs drifted past us, some like little frozen ducks, and others shaped like chairs and oversized bananas. A clump of feathers stuck to the sea surface, evidence of some recent kerfuffle. Suddenly, we found ourselves surrounded by humpback whales, one fluking every five minutes. Half a dozen surfaced within a couple hundred feet of us. One made a sweet, short, high-pitched song – trumpeting – before it disappeared beneath the waves. Just a few miles ahead of us was a landing spot at Yankee Harbour, an island of the South Shetlands that once supported a busy whaling station.

A generation ago, a boat in these waters would have been whaling, wrestling one of these ocean giants. 'Anything can change', said Steve Forrest. 'It's remarkable how well the humpbacks have recovered. And the fur seals too.' Behind us, a raft of gentoo penguins – some thirty or so – came flying through the water, playfully following the Zodiac. I thought back to a conversation I'd had with Strycker earlier that week on board the *Esperanza*. He'd been out counting for the day, at Stinker Point on Elephant Island, and returned elated, having seen two king penguins with eggs. This was the first time Strycker had seen a king penguin in Antarctica – ordinarily, they live further north on islands such as the Falklands and around the tip of Argentina. He figured they were pushing south due to climate change, but he was delighted to see them. 'These are the guys that were hunted for oil, but they came back with a vengeance', he said. 'I like to think we now appreciate penguins for what they are', he added. 'We've come a long way.'

8

GENES, DRUGS AND JUSTICE

KATE DUNCAN SAYS she often feels like Darwin with a microscope. It's an apt description for the chemist, who runs a marine drug discovery lab at the University of Strathclyde in Glasgow.[2] One of a small cohort of scientists whose work is to find new drugs from the sea, Duncan focuses on microbes, whose largely unexplored genomes are proving to be filled with gene sequences, or recipes, for chemical compounds so unique and powerful that scientists believe they may hold the clues to new medicines. 'Not many people can honestly say that their job is to discover stuff. That's really quite exciting', Duncan told me. 'Okay, these are small things, but I'm often the first person to look at the biodiversity in an area, to assess its potential.'

The potential for ocean 'biodiscovery' is mind-boggling and there are still major gaps in our knowledge. Of the ocean's complex creatures – sponges, sea lilies and the like – we've described around 10 per cent, just 230,000 of an estimated 2 million species. But the simpler life forms – the bacteria, the viruses and

2 The Duncan Lab relocated to the University of Newcastle in 2024.

also the archaea, a group of unicellular organisms that thrive in extreme environments such as anoxic deep-sea mud – make up around 95 per cent of all ocean life. Vastly abundant, they are virtually unknown to us. In each drop of seawater, there are an estimated 1 million bacteria and 10 million viruses. Placed end to end, the ocean's viruses alone would span 10 million light years. They would reach further than the closest 156 galaxies.

Duncan's microbes live in marine mud, and while some come from shallow seas around Scottish islands, others have their natural home at the ends of the Earth, in locations as far-flung as the Arctic Ocean and the Southern Ocean, deep beneath the polar ice. Often they're sampled opportunistically, by a benevolent colleague who is happy to take a minuscule sample of mud here or there while on an expedition, pop it into a glass vial, pack it with dry ice and ship it back to Scotland. It was while scuba diving in Croatia, as an eighteen-year-old backpacker, that Duncan first landed on the idea of studying the chemistry of sea creatures. 'I thought, well, wouldn't it be cool if you could look at the chemistry of these things in the ocean – things like sponges and corals.' On returning to Scotland, Duncan found that a researcher named Marcel Jaspars was running a degree course in marine chemistry at her local university in Aberdeen. 'I went to his office and I said, you know, is this a thing? Can you tell me about it? He gave me a bunch of books like Rachel Carson's *Silent Spring* to read', she recalled. Duncan signed up.

Since then, she has spent her days working on drug discovery from the ocean, both in coastal shallows and on the high seas. The microbes that pique Duncan's interest are a group of bacteria called the actinomycetes. These are among the toughest creatures on Earth. Able to survive in a huge range of environments, in lakes, inside animals, and in compost heaps, actinomycetes gather in groups of up to several million per gram of soil. Lacking obvious physical defences such as spikes or scales, they use a rich chemical arsenal to protect themselves. In the abyssal mud of the high seas, they form spores, with impenetrable walls, and enter a sleep-like state when conditions are rough. In the lab, they're easy to isolate from other microbes because you can boil them, roast them, or dehydrate them, without killing them. As other microbes die off, the actinomycetes prevail. 'They're really quite resilient', said Duncan. The most familiar are the *Streptomyces*, a group of powerful microbe killers that, in 1944, were developed into an effective antibiotic and a cure for TB. Since then, 70 per cent of all new antibiotics have come from actinomycetes. *Streptomyces* aside, there are many rare and little-known varieties, which may hold the clue to new cures. Recently, scientists have found strains from the South China Sea that are proving to be effective against cancer and another off the Canary Islands with potential as a new antibiotic.

Before setting up her own lab, Duncan had worked in Florida, Canada and California. One posting was at the prestigious Scripps Institution of Oceanography in La Jolla, San Diego, with Paul Jensen, whose work also focuses on biodiscovery from marine microbes. Part of the University of California, Scripps attracts and keeps some of the most talented marine scientists in the world. (Much like the Hotel California, once people go there, they never leave, and it's easy to see why: Scripps sits alongside of California's best surfing beaches and, just a few miles offshore, you can witness some of the best visual displays of wildlife on Earth. During a single afternoon there, I encountered schools of dolphins, shoals of tuna and marlin and, best of all, a blue whale.)

Jensen has been at Scripps since the 80s, when he moved there to work with marine chemist Bill Fenical. In 1991, Jensen and

Fenical discovered a group of bacteria, entirely new to science, from mud collected at a depth of 3,600 feet off the Bahamas. Naming them *Salinispora*, they set to work testing extracts from them against a whole catalogue of cancers. One proved especially effective. 'It killed them dead', said Fenical at the time of the discovery. The new cancer-killing agent was named salinosporamide A. It's now known as Marizomib, and is in the final phase of clinical trials to treat patients with the brain cancer glioblastoma. Recently, Duncan has discovered new strains of actinomycetes in Antarctic mud, which she is investigating for exciting new compounds – a moonshot at finding a new antibiotic or a cancer cure. This search for new medicines, for effective antibiotics but also potent antivirals and cancer drugs, is pushing scientists to explore the most extreme environments on Earth, many far offshore and deep beneath the waves. These are challenging places – variously lightless or oxygen-deprived, super-heated or at temperatures below freezing. One of the sites Jensen is currently investigating for new cures lies slap bang in the middle of the Pacific Ocean, a place he describes as being 'about as far from any landmass as you can get'. Another site, off the California coast, has been earmarked for deep-sea mining. Life thrives there and often in unusual forms.

'If you think about biodiscovery, back in the day, it was Darwin on the *Beagle*, going around the world, observing and seeing what was out there, comparing finches on one island to those on another. All of that applies to microbes', said Duncan. 'It's just that you can't see them so easily.' Just like Darwin's famous Galápagos finches, microbes evolve over time, being forced to adapt to their environs by the power of natural selection. In the extreme environment of the deep sea, there are many variables: a simple difference in temperature, ocean acidity or pressure could favour different microbes, making distinct communities. The presence of competitors or predators, which can vary hugely between different habitats and regions, also likely shapes the communities that thrive. 'Just a single sand grain can have thousands of different microbes attached to it. And if you go a couple of sand grains down, is it going to be the same thousands attached? No one knows', said Jensen. 'This is a big frontier, and we haven't really scratched the surface of it.'

ALEXANDER FLEMING SAW the value in microbes. In 1928, his chance discovery of penicillin in an unwashed Petri dish led to the development of the first antibiotic. Although he immediately realised that penicillin could kill *Staphylococcus* bacteria, and thus save countless lives, it took the pressures of a world war to convert his discovery into a mass-produced medicine ready to be deployed with the troops. What followed was the golden age of antibiotic discovery, a period that lasted until around 1960, when most of the antibiotics still in use today were found. The majority had their origin in soil bacteria, and many – such as Streptomycin and Vancomycin – are now household names. But it's been over thirty-five years since the discovery of a new class of antibiotics. In the intervening years, research has been hampered by tight regulations and lack of funding, with few new candidates entering the drug development pipeline. Meanwhile, the microbes we routinely treat with antibiotics have been developing resistance to existing varieties, giving rise to an alarming string of antibiotic-resistant infections. The 'discovery void' of the past thirty-five years poses a threat that could, in theory, claim 10 million lives each year by 2050. The problem is that without drugs to treat bacterial infections, a simple hospital procedure, such as a knee replacement, becomes life-threatening.

Despite Fleming's success with penicillin, the pioneers of biodiscovery didn't start their search with microbes. Amazingly, their work began as a seamless continuation of colonialism, whereby wealthy 'bioprospectors' explored nature for new products, usually in the most species-rich, cash-poor parts of the world. As a method of drug discovery, this has been extremely effective, if rightly controversial for the fact that it has involved the theft of national biodiversity and Indigenous knowledge. Nevertheless, it has given rise to numerous treatments, including quinine, the anti-malarial medicine discovered from cinchona bark, and aspirin, a painkiller derived from the bark of the willow tree. It wasn't until the development in 1942 of the aqualung, however, that the search for ocean-derived drugs began in earnest. Suddenly, the shallow seas were a new frontier; scientists painstakingly scoured the world's coastlines to find new species, which they would collect, grind up and analyse for potent chemical compounds, testing them against all sorts of infectious agents.

One of the earliest breakthroughs was made by a chemist called Werner Bergmann. During a dive off the Florida coast in 1945, he came across a seemingly unremarkable flat brown sponge, later named *Cryptotethya crypta,* from which he isolated two new compounds. Almost forty years later, these compounds led to the first approved therapy for human immunodeficiency virus (HIV), a treatment trademarked as AZT by Burroughs Wellcome. To date, eighteen drugs have been successfully developed from marine organisms found within national waters. Yondelis, used to treat ovarian cancer, was derived from an inconspicuous sea squirt, and Prialt, an effective painkiller, has its origins in a marine cone snail with a lethal venom. The first treatment approved for use against Covid-19, in 2020, named Remdesivir, came from the same compounds that gave rise to AZT. Halaven, a blockbuster anti-cancer drug derived from a Japanese sea sponge, has annual sales in the range of US$340 million to $680 million.

Since the 1950s, researchers have discovered almost 34,000 marine compounds with commercial potential, for a wide variety of uses. About 250 of these are therapeutic products at the pre-clinical stage, but the applications are wide-ranging: an antifreeze protein from the Arctic cod, for instance, has been used to improve the texture of ice cream. An enzyme extracted from a microbe along the Mid-Atlantic Ridge has been used to develop a biofuel. Other enzymes, extracted from a bacterium found on a hydrothermal vent, have been used to develop virus tests, including those used to diagnose Covid-19. Genes from marine organisms are also used in food stabilisers, in laundry detergents and to produce low-emissions feed for cattle. The marine biotechnology industry is projected to reach $6.4 billion by 2025 and scientists in nations with advanced research programmes, like Duncan, are now looking to the unexplored genomes of high-seas organisms in the hope of finding new leads.

The early pioneers, such as Werner Bergmann, who discovered the sponge that led to AZT, were avid collectors of ocean life. Another was Paul Scheuer, an organic chemist at the University of Hawaii at Manoa, who is now regarded as the founder of biodiscovery research. Scheuer first came to the US as a refugee from Nazi Germany, having been denied the right to a university education under Hitler's Third Reich. Starting as a tanner's apprentice in the small Hungarian town of Pécs, he immigrated to the US, won a scholarship to Harvard graduate school and, in June 1950, was offered his first job as an assistant professor at the University of Hawaii. Despite what he described as a 'profound ignorance' of both the Hawaiian islands and the university – having never visited either – he and his then fiancée, Alice Dash, grabbed the opportunity. On 5 September 1950, the pair were married in Harvard chapel and the following day they left for Hawaii.

Scheuer spent the following years building up the university's programme on the chemistry of natural products. He began his investigations by exploring the islands' plants, including the native 'awa' – a powerful narcotic, renowned for its ceremonial uses – but he soon moved on to explore the creatures in the surrounding seas. 'An even more dramatic experience than Polynesian rainforests was the ocean – warm, blue and rich in animals and plants of which I was totally ignorant', said Scheuer, in an interview decades later. Like Bergmann, Scheuer collected an astounding variety of animals and plants for analysis. Most were immobile creatures that lacked tough skin, visible spines, teeth or claws. Yet they survived incredibly well on the seafloor or the rocks to which they clung, leading Scheuer to suspect that they harboured a rich chemical defence arsenal that might be harnessed.

The first evidence that confirmed his suspicions came when a friend, a zoology student named Bob Johannes, added a sea slug to his home aquarium. Known as the fried egg sea slug, *Phyllidia varicosa* is a brilliantly coloured nudibranch – navy, white and bright blue with yellow splodges. Johannes' delight at acquiring the vibrant creature soon vanished when the tank's other inhabitants – his pet fish and shrimp – died. Suspecting the new arrival was responsible, Johannes traced the cause of death to a mysterious substance oozing from the sea slug's back. Fascinated by his friend's story, Scheuer was desperate to study the substance further, but getting his hands on the sea slugs proved tricky. He sent word out at the university that he was looking for some samples, and went as far as to place an advert in the local malacology (mollusc-expert) newsletter. One day he received a call from the nearby island of Maui, reporting a sighting of the slugs swimming at a place called Maalaea Bay. Two co-workers flew to Maui to collect the goods, returning a few days later laden only with bags of Maui's famous potato chips. The slugs had vanished. At last, in the summer of 1973, a fellow chemist Jay Burreson spotted one of the slugs with its bright colouration swimming at Pupukea Bay, on Oahu's north shore. Burreson captured the animal and, with some simple experiments, was able to verify Johannes' initial observation: the sea slugs were toxic.

But he also made another, even more important, discovery. Astonishingly, after just a few days in captivity, the sea slug's toxicity waned. Another year passed before he found an explanation: during a dive in Pupukea Bay, he noticed the same species of sea slugs feeding on an off-white coloured sponge. The lethal chemicals were harboured in the sponge, and only on eating the sponge did the sea slug develop its own defences. Without its noxious diet, *Phyllidia varicosa* was harmless. The toxin was pupukeanane, a potent compound that Scheuer helped develop into numerous commercial products including anti-malarial formulas and anti-fouling, which is used to keep the hull of boats clean. Another discovery from Scheuer's lab was Kahalalide F, a chemical compound unexpectedly found in a different sea slug, *Elysia rufescens* (which means 'becoming red'). Kahalalide F has been the focus of clinical trials as a treatment for lung and colon cancer and also for the skin condition psoriasis. Its use is now licensed to various pharma companies including Spanish PharmaMar and US Medimetriks.

What Burreson had discovered was that toxins can be acquired by predators from their prey. What he didn't yet know was that, in most cases, the prey didn't actually produce the toxins. Years would pass before the rest of the story came to light, and when it did, it came from an unlikely source – a rove beetle found deep in the Brazilian jungle. With its metallic blue-green wings and bright orange abdomen, the rove beetle's colouring offers a clear warning to any creature who dares come too close. It does this because it harbours a toxin, called pederin, which is more potent than cobra venom. Brush against, or crush, a rove beetle, as you might easily do on a trek through the jungle, and the resulting blisters can leave you with lasting scars. For years, scientists assumed that the beetle itself produced this nasty toxin, but in 2001 a German scientist named Rupert Kellner discovered that the toxin was made by a microbe that lives in the beetle's gut. At that stage, pederin had also been found in creatures elsewhere, including a deep-water sponge off the coast of New Zealand. It took a few more years for another scientist to find the genes that coded for the toxin in the sponge. They were bacterial genes. It soon became clear that sponges, beetles and lichens all had the same toxin, produced for them by symbiotic microbes. The discovery made scientists realise that many of the potent compounds they'd found in large marine invertebrates, and had in turn developed into successful drugs, may well have been produced by the microbes that live in harmony with them. Before long, scientists started to turn their attention to microbes.

IN 2006, AN American biotechnologist and entrepreneur named Craig Venter dreamed up an ambitious idea: to gene sequence not just one microbe, but the ocean's microbes. A few years earlier, he had succeeded at sequencing the first human genome – his own – beating a parallel US government–funded effort. The award-winning achievement took him eight years and cost $2.7 billion. Yet Venter wasn't satisfied with his contribution to the genomic revolution. And so he began a voyage, inspired by Charles Darwin's journey on the *Beagle*, across the high seas. Starting in the Sargasso Sea, Venter took a 95-foot sloop named *Sorcerer II* across to Halifax in Canada, down the US east coast, to the Caribbean Sea, through the Panama Canal, to the Galápagos, across the Pacific, and over to the Indian Ocean.

He spent two years at sea covering 32,000 nautical miles, collecting and sequencing marine microbial genomes along the way. His methods were incredibly simple: his team sampled seawater and put it through a series of increasingly fine filters to collect different-sized organisms. Back on land, they sequenced the genomes of every microbe they'd collected, working at a rate of 100 million letters of genetic code every twenty-four hours. By venturing into uncharted waters, Venter sequenced millions of genes, many previously unknown. In one barrel of seawater collected from the Sargasso Sea, for example, they discovered 1.3 million new genes and 50,000 new species. But Venter's ultimate aim was vastly more ambitious: from the genomic data, he planned on creating the world's first artificial organism. In 2010, he succeeded, building an entirely synthetic microbe. 'It's a living species now', Venter said at the time. 'Part of our planet's inventory of life.'

More than any invention since the microscope – which allowed Fleming his groundbreaking discovery – gene sequencing has revolutionised our ability to see, and study, the microbial world. It has also turned the field of marine biodiscovery on its head, from a world of plundering nature to one of tinkering with nature. 'When the field of marine natural products took off, people were collecting big easy things. And there's been a lot of success stories there', said Paul Jensen, the marine microbiologist at Scripps. Usually, just small amounts of samples were needed. Still, said Jensen, 'There were ethical issues over collecting these kinds of things. Most of the time, it was minimal, but there were some egregious cases, in the early days, of people collecting large amounts of animals.' While Jensen describes microbes as a renewable resource – so abundant that no harm comes from sampling them – scientists needn't always harvest wild populations. Instead, they interrogate species' genomes to find sequences that code for useful chemistry.

The method Venter used was known as shotgun sequencing; now standard practice, this involves splitting DNA randomly into small fragments that can be sequenced by a machine; once the sequencing is complete, a computer programme looks for overlap between fragments and stitches the whole genome together in the correct order. Voila: a DNA extract from a sponge or a microbe results in a genome sequence that can be read and stored digitally. Thanks to advances in genomic technologies, it's now possible for machines to read genomes – the entire set of DNA instructions for an organism – at lightning speed. A human genome can now be sequenced in an hour, for as little as $100. As well as having the tools to read genomes, we also have tools to manipulate them: we can cut the bits that we like out of one genome and paste them into another; we can also write entirely new genetic sequences, to create new products, technologies and even new life forms.

More often than not, when scientists take a genetic sequence from a marine animal, plant or microbe, it is for a similar use elsewhere. Take the Arctic cod, for instance, a fish species that can survive at temperatures as cold as 28.7°F without its blood freezing. Scientists discovered that it contains a gene sequence that specifies a protein, which binds to ice crystals, preventing its cells from freezing. They have since taken this sequence and used it to make smooth ice cream, as well as to store sperm safely in donor banks. Or, consider the genetic sequence that codes for bioluminescence in the jellyfish *Aequorea victoria*, which has been used to create a neon cat – this was done not just for fun, but to better understand how feline AIDS develops.

This means, of course, that genetic resources found from the high seas, such as Venter's or Duncan's microbes, are no longer just physical specimens but also gene sequence data uploaded to online repositories. The most well-known of these is GenBank, an open-access, annotated collection of all publicly available gene sequences and the proteins they encode. From GenBank, sequences can be downloaded, synthesised in a lab and used to develop a new product. Increasingly, this 'digital sequence information' (DSI) is all that's needed to create and mass-produce a new drug. For example, Kahalalide F, the compound isolated from a sea slug and currently being tested against cancers and psoriasis, is now created synthetically, from digital information, in a lab. When the Spanish company PharmaMar licensed it to US Medimetriks, the firm that is testing it for psoriasis, all Medimetriks needed was the sequence data, just like a readout of computer code. 'Over time, the focus has moved from collecting a zebra or a starfish to collecting a tiny little sample of that thing to collecting just the genetic sequence data. At this point, you may not even need the zebra', said Robert Blasiak, an ocean governance researcher at the Stockholm Resilience Centre. 'That's basically where the whole industry has been moving for the past twenty years.' The development of digital sequence information has huge advantages: it requires no harvesting, and its commercial potential is virtually limitless, as you can sell a gene sequence – essentially the equivalent of a blueprint or a recipe for creating it – online multiple times. Digital sequence information now holds the biggest opportunity for commercialising the ocean's genetic resources.

IT'S BEEN AN exciting time for biodiscovery scientists, but also an anxious one. Though biodiscovery has been successful, it has also been controversial, often associated with commercially driven 'bioprospecting' and sometimes even 'biopiracy', a term that describes the misappropriation of native plants and animals and local knowledge for commercial gain. It wasn't until 1993 that a Canadian environmentalist, Pat Roy Mooney, coined the term biopiracy to name what he saw as an alarming trend: western institutes and corporations taking, and patenting, biological and genetic resources from developing nations. In biodiversity hotspots such as the Amazon rainforest, the problem was rife. From tree rubber to acai berries, the region's resources were being turned into a vast array of products, worth billions to overseas companies. Yet local farmers were reaping few, if any, of the profits. Around the same time, the Indian government found itself in the somewhat absurd position of having to fight off a US company looking to patent the spice turmeric for medicinal purposes. In South Africa western companies were developing a range of products from local plants without offering to compensate the KhoiSan communities who had used and cultivated them over thousands of years.

That was why, in 1995, Mooney formed a group, the Coalition Against Biopiracy, to lobby world leaders to take biopiracy more seriously. One of their strategies was to name and shame the corporations, countries and institutions misappropriating biological or genetic property. No one was beyond reproach. The worst offenders were given a Captain Hook Award, while those defending local communities against the threat of biopiracy were given a Cog Award, named after a type of tenth-century wooden sailing ship that proved useful in repelling pirate attacks. Some of the accused were obvious targets for Mooney and his peers:

Monsanto, for filing a sweeping patent on soybeans, and the US Patent and Trademark Office, for attempting to assert IP rights over Ayahuasca, the Amazonian medicinal hallucinogen. Some were perhaps less obvious. In 2006, the University of California, Davis, was named for patenting a blight-resistant rice gene originally from Mali. That same year, Craig Venter was singled out as the world's 'greediest biopirate' for attempting to sequence the ocean's microbes.

Though Venter posted the data on GenBank, free for anyone to use, and invited scientists from developing countries to collaborate, Mooney and his allies worried that Venter was paving the way for the world's wealthy to patent genes and even life forms. Eventually, in 2014, the UN passed the Nagoya Protocol, a law intended to prevent biopiracy on land and in national waters, making it illegal for a company or a researcher to collect any biological specimen – a plant, an animal, even a microbe – from a foreign country without a permit. In addition, any product patented using Indigenous resources or knowledge would need to compensate the original stewards or knowledge-holders. So far, this has only resulted in one successful deal, agreed in 2019, for a portion of the revenue from sales of rooibos tea to go to the KhoiSan communities of South Africa as compensation for use of their traditional plants and knowledge. The deal will see South Africa's rooibos tea industry pay an annual levy of 1.5 per cent of the price of raw unprocessed rooibos (*Aspalathus linearis*), a variety endemic to the Cederberg region, north of Cape Town, into a trust for local communities. In July 2022, the rooibos industry made its first payment into the fund of approximately $715,000, split fifty-fifty between the San and Khoi communities.

Even after the Nagoya Protocol came into force, the genetic resources of the high seas were up for grabs. Under the United

Nations Convention on the Law of the Sea (UNCLOS), which came into force in 1994, nations have the right to extract and profit from fisheries in international waters, but they must share the profits from minerals – such as cobalt – extracted from the international seabed, which is seen as the 'common heritage of mankind'. The Law of the Sea, however, said nothing about how nations should share high-seas genetic resources. 'Some of the most exciting stuff that the pharmaceutical industry and the chemical industry are interested in is probably down there – in ocean vents and underneath that Arctic ice floes. That's where the neat stuff is', said Mooney, who is now in his seventies, his slightly cherubic face framed by tufts of white hair. Agreeing a law to regulate their use has been tricky, in part because the genomes of the high seas, Earth's largest commons, have arguably been owned by either no one or everyone.

Historically, the EU, and nations including the UK, US and Japan, with the technology and financing to scour the deep sea in search of new products, have argued for the right to patent, and solely profit from these resources. They have wanted to maintain the liberal access regime of the 'free seas'. Developing nations, typified by African and Caribbean nations, have argued that profits, and other benefits such as data derived from the use of these genetic resources, should be shared with all nations. Part of their motivation is that they have felt left out of this area of research and commerce and, seeing the opportunities ahead, have been desperate to become involved. They've argued that genetic resources are common heritage, and so commercialising them without compensating everyone on Earth (even those in landlocked countries) amounts to biopiracy. 'It's not the same kind of biopiracy as on land where we can clearly say that they're stealing from humans or Indigenous Peoples, because I don't know that anyone's claiming ownership of an ocean vent yet', said Mooney. All the same, he has worried that corporations ultimately want to find these resources, patent and own them. Mooney's view is that while early explorers and colonists may have misappropriated plants and animals, it was the introduction of intellectual property rights that allowed companies to profit hugely from their discoveries. The patenting system has meant that a company can develop a blockbuster drug from a high-seas genetic resource and become vastly wealthy in the process – the extent to which they would need to compensate others has been a source of tension between rich and poor nations.

Part of the problem is the growing inequality between countries that can afford to explore and use marine genetic resources. In the past three decades, ocean exploration has largely been a luxury afforded to wealthy nations, most notably the US, UK, Canada, Japan, Australia, Russia and Germany. Currently, 90 per cent of patents on natural products are owned by corporations or institutions in the global north, with 90 per cent of the source material coming from the global south. If you consider, for a moment, that the cost of a research expedition can run to several hundreds of thousands of dollars, and millions in some cases, notwithstanding the investment required to build icebreakers or to develop new technologies such as deep-sea submersibles and sophisticated lab equipment, it's clear that only very few nations can afford to be involved.

On the high seas, far offshore, the costs are higher still. And the fact is that marine drugs remain horribly overpriced, taken as evidence that wealthy corporations are benefiting hugely from their discovery, while few others do. When Gilead Sciences received FDA approval for its antiviral Remdesivir as the first Covid treatment in May 2020, it priced a five-day course at

$2,340 for governments in the developed world, and $3,120 for US insurers, including those on Medicare and Medicaid. Meanwhile, an independent watchdog valued the drug (based on how likely it was to reduce patients' hospital stays) at $310 per course, one-tenth its actual price.

Complicating matters further is that the nature of what's being misappropriated has changed. One outcome of the genomics revolution is that most innovation comes from digital sequence information – the instructions for making a specific compound or even a life form. 'With the accelerating tools of genome-editing and synthetic biology, today's biopirates no longer need to carry their booty offshore in boats and airplanes', said the Master of Ceremonies at the last 'Captain Hook' shaming event. 'They can upload DNA as digital sequences in one location and then recreate it as synthetic DNA on the other side of the planet', he said. Indeed, there is some evidence, from a 2018 study by Robert Blasiak at the Stockholm Resilience Centre, that corporations have taken advantage of this trend in digitisation of genomes, and their lack of clear ownership, to register, and patent, marine gene sequences. Searching through 7.3 million genetic sequences on GenBank, the data repository to which Craig Venter and also Kate Duncan upload their sequence data, Blasiak and his colleagues identified roughly 13,000 sequences with marine origins. Of these, half had been acquired by a single company, the German conglomerate BASF, which calls itself 'the largest chemical producer in the world'.

This figure was seized upon by the media as a sure sign of growing corporate control over the ocean's genetic resources, even though the numbers involved are incredibly small and not all of these patents will even become commercial products. Still, it's worth noting what these sequences code for and why they might interest a corporation such as BASF. Several of the sequences registered by BASF derive from the deep-sea Pompeii worm (*Alvinella pompejana*), named for its ability to survive the extreme heat of hydrothermal vents. A useful addition to skin creams, its genes now appear in eighteen patents registered to BASF and to a French research institution. In another example, BASF has decoded the genetic sequence responsible for producing omega-3 in a marine microbe and has spliced the sequence into a rapeseed plant so that it can produce omega-3–enriched rapeseed oil. All going to plan, the fortified rapeseed oil will soon be available for human consumption. If we can figure out the genetic make-up of other organisms that thrive in the unusual heat, darkness, pressure or acidity of high-seas environments, they too may be commercially exploited to create a whole variety of novel products and blockbuster cures.

'I'VE BEEN CALLED a biopirate', said David Newman, who, for around forty years, was chief of the natural products branch at the US National Cancer Institute – the largest and best funded biodiscovery programme in history. 'Anytime we were somewhere foreign, we'd bring a sterile tongue depressor and a sterile wallet, just so we could take a sample of soil here or there when we were out having a beer or taking a pee in the wilderness. It was just what we did – we took samples of biodiversity where we could.' But, he said, 'That was before the new laws came in.' When the Nagoya Protocol came into effect – to prevent biopiracy on land and in coastal waters – it had a profound impact on research. Although it was a global law, it was up to countries to implement it in a way that made sense to their situation. Some, such as Ecuador, Colombia and Sri Lanka, imposed strict regulations on collecting any natural specimens, even for basic research. They were fearful of biopiracy. Suddenly, it became incredibly difficult to access nature on land and in the coastal waters of these countries. Navigating the various administrative requirements and bureaucratic hurdles was such a strain that many researchers changed tack. Cruises were called off. Plans redrawn. Funding applications altered. The whole experience was bruising for science.

Nations became wary of implementing similar rules for the high seas. The attempt to find a resolution to this seemingly intractable problem has taken two decades of negotiations. At the heart of these discussions have been difficult questions such as: are gene sequences from the high seas shared resources that must benefit all of humanity? If so, where does misappropriation start and end? Would it be an act of 'biopiracy' to use online data to synthesise a gene sequence or to use archived material – DNA from a museum specimen, for instance – in developing a new product without sharing the profits of commercialisation? What if the final product was mostly synthetic with one gene from a marine fish or sea slug that was harvested long ago, before any such rules were in place? There's also been the question of whether, fearing biopiracy, we should regulate access to information held as online data. Developing fair rules for this fast-moving industry is complex: imagine, for a moment, how difficult it would be to trace the origin of a digital gene sequence once it has been synthesised and incorporated into a product; even more so again when a product is designed using genes from different organisms collected in various jurisdictions. As the search for new marine drugs moves further offshore into the unclaimed ocean, these questions have become more complex.

Resolving these dilemmas is more urgent than it might seem. In 2020, ocean governance researcher Robert Blasiak authored a study in the journal *Nature Sustainability*, in which he coined a new term: the 'ocean genome', which describes the wealth of genetic material held in all marine life forms, including both the physical genes they possess and the digital information they encode. Blasiak and his peers warned that the ocean genome is under threat. As we lose marine biodiversity – from climate change, pollution, overfishing and the encroachment of new industries, such as deep-sea mining, into the high seas – we also stand to lose these genetic resources and the potential recipes they contain.

Take, as an example, the scaly-foot snail (*Chrysomallon squamiferum*), which in 2019 became the first species officially endangered by the threat of deep-sea mining. The rare mollusc, which now features in the International Union for Conservation of Nature (IUCN) 'Red List' of species under threat, occupies just three known locations in the deep sea – all hydrothermal vent systems earmarked for seabed mining. But the scaly-foot snail is more than just a rare creature: it is unlike any other known, building its shell from iron sulphide and covering its soft, fleshy foot with iron plates, all for its own defence. Scientists have figured out that it's actually a microbial symbiont that builds the snail's armour and, in 2020, another group of scientists decoded its genome, in the hope of gaining insights that might allow us to develop better human armour. As extractive industries progress offshore, you have to wonder how many discoveries, with amazing applications, will be lost to us.

In contrast to deep-sea mining, exploring the ocean genome can be a lucrative, sustainable and equitable enterprise. It could genuinely enrich humanity, by allowing us to develop a harmless ocean-based industry, all while providing us with the formulas for drugs to treat even the most deadly diseases. If we choose to harness the value in the ocean genome, and to share it fairly, would we need to consider destroying ocean habitats with mining, or extracting the last bit of profitability from our fisheries? Forty years ago, the international seabed was declared the common heritage of humankind. The hope was that this arrangement would bring a sustainable industry to the ocean, in the form of seabed mining, and equity to the world, in the form of shared profits. Now, it's evident that neither are likely. But with ocean genomics, we now have another chance to get it right.

In March 2023, nations agreed, for the first time in history, that high-seas genetic resources are common heritage. This means that data from marine genetic resources must be made openly available, opportunities must be given to scientists from developing countries to partake in this research, and the spoils of any commercialisation must be shared. That includes profits from product development, a portion of which will be channelled into a global fund for ocean protection. The new law – which will take several years to come into force – regulates for both physical samples and digital sequence information. It is the strongest international agreement ever to counter the threat of biopiracy. Will this solve the problem? That remains to be seen. The US, having not acceded to the Law of the Sea, is unlikely to join the agreement, though it may play by the rules. As for the Nagoya Protocol, nations will have to implement their own rules to comply with the law; some will be laxer than others, favouring corporations. And as genetic engineering evolves, it will throw up new moral dilemmas. One, for instance, is the likely growth of 'bio-inspired' products, some of which will be created using artificial intelligence, and based on the ocean genome, rather than actually utilising it.

Meanwhile, the reality, of course, is that few countries have the capacity to harness the commercial potential in marine life forms. The road to developing a new drug is still long and arduous. With all of the advances in genomics, it may sound as though the pace of drug discovery has quickened. For every 5,000 new candidate drugs, only 1 makes it to market. And on average, it takes twelve years to develop a new drug, or eight to ten years for an antiviral. But it can be longer: it's already been almost twenty years since Paul Jensen and Bill Fenical first reported their discovery of the cancer-killing agent salinosporamide A. In 2020, Bristol Myers Squibb began phase 3 clinical trials to test it against the brain cancer glioblastoma. At the time of writing, the trials are still ongoing.

In the last two decades, all big pharma companies have closed their biodiscovery branches; the preferred model now is that they buy the patent for a product once it's at, or past, the pre-clinical stage. That way, they avoid a lot of the initial outlay. The majority of biodiscovery is now carried out in research institutes and universities by scientists, such as Kate Duncan, who stand to gain little beyond peer recognition. Duncan described drug discovery as 'really tough'. 'It's easy to feel doom and gloom about antibiotic discovery', she said. 'There are so many hurdles – like the approval process and the timeframe of clinical trials. But you have to be hopeful and active in this area.' For many who work in drug discovery, the term biopiracy has been highly charged. Duncan visibly bristled when I mentioned it. For her, bioprospecting is enough of a misnomer. 'This is about discovery', she said. 'Nature is a better chemist than we will ever be. We're just utilising that.'

9

DEEP TROUBLE

THE TWENTY-SECOND OF March 2001 saw a scene unlike any before. The lucky few gazing skyward from Fiji – the best viewpoint on land – watched mesmerised as a procession of bright golden lights tore across the night sky, leaving in their wake a long shimmering trail. Closer to the action were the passengers who had paid US$1.8 million to witness the event aboard two planes. Out at sea, unsuspecting fishers heard the sonic boom before clocking the mass of debris hurtling downwards. Finally, after a journey of some 1.8 billion miles, the remnants of Mir, the Soviet space station, splashed down into the South Pacific Ocean at 17:58 local time. Decommissioned after fifteen years in orbit, most of the 165-ton spacecraft burned up as it re-entered Earth's atmosphere. The remainder fragmented into chunks – approximately 1,500 pieces in total, some the size of a small car – and met a watery end in the Pacific Ocean, loosely scattered around Point Nemo.

Roughly 3,000 miles off the east coast of New Zealand and 1,800 miles north of Antarctica, Point Nemo is, quite literally, the furthest ocean point from land. Point Nemo wasn't discovered in the traditional sense; rather it was mapped in 1982 by a Canadian Croatian survey engineer, Hrvoje Lukatela, using a programme he designed called Hipparchus to locate the most remote part of the high seas. Lukatela calculated the coordinates of three equidistant coastal spots that each had Point Nemo as their furthest ocean location – Ducie Island, one of the Pitcairn Islands, to the north; Motu Nui, a tiny dot of land off Easter Island, to the northeast; and Maher Island, off the Antarctic coastline, to the south. Once described as 'a load of water, surrounded by even more water', Point Nemo is known to scientists as the 'Oceanic Pole of Inaccessibility'; to space agencies, it is the 'South Pacific Ocean Uninhabited Area' or SPOUA. Its very isolation, however, has made it famous for another reason: it's now the world's largest spacecraft cemetery, the place where space agencies and private companies purposefully crash spent rockets, satellites and decommissioned space stations. Since 1971, more than 265 space objects have been sent, hurtling back through our atmosphere, to this remote spot. That they've met their fate in this obscure location is, for the space industry, a triumph.

At this very moment, millions of human-made objects encircle our planet. The largest of these is the International Space Station (ISS), which weighs about a million pounds and, measuring 355 feet end to end, is roughly the size of a football field. Most space objects are, however, debris – the nuts and bolts of worn-out rockets, the shards of old solar panels and flecks of paint which have, through some cosmic altercation or another, been violently ripped from the metal structures they once coated. In space, the threat posed by debris is all too real. In July 2015, a piece of metal that broke off a Soviet military weather satellite – named Object No. 36912 and roughly the size of a dinner plate – came within 1.5 miles of the ISS, triggering an emergency response from ground control in Houston. Had it made contact, it could have annihilated the space station: travelling at thousands of miles an hour, a piece of space debris the size of a bolt can explode, upon impact, with the force of a hand grenade. Even flecks of paint can seriously damage spacecraft. Each celestial collision, of course, forms more debris, and the more objects we send into space, the more collisions happen. According to an analysis published in January 2022, there are currently about 4,550 satellites orbiting Earth (although the Union of Concerned Scientists, which keeps a tally of operational satellites, puts the number closer to 6,500). Of these around half are for communications purposes, and 36 per cent are owned by Elon Musk's SpaceX, which is looking to launch upwards of 1,000 new satellites into orbit every year with their Starlink programme. Amazingly, by some estimates, there may be as many as 100,000 satellites orbiting Earth by 2030, posing a clear danger to future space missions.

'There is so much junk that we are worried one tiny collision could trigger a big chain reaction', a NASA spokesperson said. Eventually there could be so many collisions that low Earth orbit would become unusable, clogged with trash. To prevent that from happening, the aerospace industry either blasts its old satellites into a higher orbit or plans for a 'controlled re-entry', sending them hurtling back towards Earth and, more specifically, the high seas. The friction of re-entering the Earth's atmosphere is a useful form of waste disposal: small objects – shoebox-sized satellites and the like – burn up on meeting the Earth's atmosphere, but larger ones – such as rockets and space stations – make it, somewhat intact, to a watery grave in the deep ocean. By locating their spacecraft cemetery at the point furthest from land, space agencies can keep within the acceptable safety limits established by the aerospace industry.

Mir's re-entry was, by all accounts, successful. In January 2001, two months before splashdown, the Soviets sent a formal note to the UN secretary-general informing him of their plans for decommissioning the spacecraft in the South Pacific. The note estimated that an 'aggregate mass of unburnt structural elements' would reach the Earth's surface; in total, about 27 tons of space junk. For some, it was a suitably distinguished end to a station that had hosted 104 astronauts over its lifespan and that, aside from the Moon, was the heaviest object ever to orbit the Earth. Mir had survived the collapse of the Soviet Union in 1991 and had helped ease the end of the Cold War by hosting US astronauts. But Mir's decline had begun, signalled by general wear and tear, and hastened by larger accidents and breakdowns. In 1997, for instance, its crew were confronted with a fire on board, after an oxygen-generating canister burst into flames. Just a few months later, Mir became the target of the worst collision in space history when an uncrewed cargo craft, loaded with food and fuel, slammed into one of the station's linked nodes, inflicting a puncture wound that threatened the cosmonauts' air supply. The crew survived, but the crash took its toll on Mir. It was time to retire the Soviet station.

On 22 March 2001, Mir, emptied of its crew, was carefully repositioned from a circular to elliptical orbit, a move that would ready it for the plunge. Shortly after midnight local time in Moscow, Russia, a cargo ship laden with fuel initiated the final braking manoeuvre. At ground control, nerves ran high. Chuffing cigarettes and eyeballing a large map on a screen tracking Mir, mission controllers busily plotted the likely course of the satellite during its final orbital laps, 135 miles above the Earth. A little over five hours later, around 33 tons of Mir, roughly as predicted, hit the surface of the dark ocean. What remained was a mass of pressurised tanks, sturdy enough to survive space and a fiery descent into Earth's atmosphere. Now, resisting the crushing depths of the South Pacific, the remnants of Mir would start a new life in the deep sea.

But not everyone was pleased. A club of Latin American nations known as the Rio Group had become worried that allowing so-called splashdowns in the ocean, without any thought to the impact, would 'turn this common province into a dumping site for dangerous materials'. Though Mir wasn't the first space object to end its working life in the Pacific (Salyut 1, a smaller Russian space station, had previously met the same fate), it was the largest spacecraft to have ever existed, and as such the largest to have re-entered Earth's atmosphere. The Rio Group voiced their concerns to the UN, but with Russia opposing any legal restrictions on the practice of disposing space waste at sea, the splashdown went ahead. It set a precedent. Since then, Mir has been joined by defunct satellites, rocket parts and an automated robot, the Jules Verne, that delivered cargo to the International Space Station.

It now seems certain that Mir's successor, ISS – which is coming to the end of its useful life – will soon make the same journey. A beast of many parts, ISS is made up of a series of pressurised modules filled with equipment weighing in total around a million pounds. On the outside, there are hundreds of thousands of solar cells; on the inside, there are labs, a gymnasium and a living area comparable to a six-bedroom house, which has been described as 'humanity's most expensive real estate'. It's hard to know what of this will survive re-entry and end up in the deep sea.

It's not only Point Nemo that's filling with space debris; it's other parts of the high seas too. Some space objects effectively decommission themselves; they fail, fall out of orbit and begin what is known as an 'uncontrolled re-entry', where the force of gravity pulls them towards Earth. In May 2021, for instance, the upper stage of a rocket that had carried part of China's space station into orbit re-entered Earth's atmosphere and splashed down into the Indian Ocean, just north of the Maldives. A year earlier, another rocket upper stage made a chaotic return to Earth, spraying shrapnel into the sea off the west coast of Africa. The unplanned re-entry sent fragments far and wide: pipes landed on a house in the Ivory Coast and other debris reportedly wrecked a cheese-making machine in Mahounou village in Bocanda, Côte d'Ivoire.

'Taken together, the controlled and uncontrolled re-entries happen a lot. So we're talking daily, essentially', said Stijn Lemmens, who researches space debris with the European Space Agency in Paris. 'For large objects, like a big satellite or an upper stage, we're talking about once every one or two weeks', he told me. So on any given day of the year, a piece of space junk is literally tumbling out of the sky towards our earthly home. Since Sputnik was launched in 1957, around 24,000 space objects have re-entered Earth's atmosphere, a number that is only set to grow as the space industry expands. It's thought that roughly three-quarters of re-entries lead to a splashdown and that somewhere between 10–40 per cent of the original material survives. Most of this debris is small and because 70 per cent of the Earth is ocean, it usually ends up there, which means that we don't even notice it. Yet in solving the problem of space debris, we have allowed an exponentially growing industry to dispose, without restriction, of its junk in the half of our planet that is unclaimed ocean.

In May 2021, when the Chinese rocket stage CZ-5B made its frenzied fall towards Earth, the fear wasn't just that it would hit someone, but that it would also contain toxic materials.

'There is potential danger to marine life here. There are a number of substances and matter that can affect vulnerable ecosystems', said Vito De Lucia, a legal theorist at the Arctic University of Norway in Tromsø. Mingled in with the hard components – the metal structures and so on – are chemicals, including hydrazine, used as a propellant in rockets and highly toxic to water organisms. There are clues, from unplanned touchdowns on land, that such toxins do survive re-entry: several rocket boosters, for instance, have fallen near populated areas in China, one close to a primary school and another in Kazakhstan, releasing clouds of an orange toxin known as a BFRC (Big F***ing Red Cloud). In the deep ocean, such breaches go unseen. But as with other pollutants in the ocean – such as debris from mining – currents can carry toxins and other harmful substances far beyond the disposal site, where they can infiltrate food webs, with potentially drastic impacts on the wider marine ecosystem – and beyond.

Under the right conditions, even living organisms can survive re-entry, bringing new life forms to the deep sea. The nuclear reactors, often used as a reliable power source for extended space missions, also pose a serious risk. When the lunar module of Apollo 13 re-entered Earth's atmosphere in April 1970, it brought with it a nuclear battery. A source of plutonium, this battery ended up in the South Pacific, near the Tonga Trench. There, it will remain radioactive for the next 2,000 years.

Is any of this especially harmful? The truth is that no one knows because almost no one has assessed the likely impacts of splashdowns on marine life. During heated exchanges between the Rio Group and the Russians, in 2001, on whether Mir should be retired to the South Pacific, the Russians argued that lack of knowledge suggested no need for concern. That's not, however, how we usually gauge whether an activity is environmentally acceptable. 'Normally, in an environmental impact assessment, you would look at a number of scenarios, including the analysis of cumulative effects, and take a precautionary outlook', said De Lucia. 'Even though an individual re-entry may not be problematic, in the context of a degrading environment experiencing several pressures, you may come up with a different answer.'

The fact that a part of the ocean is filling up with rockets' secondary payloads, mini satellites, and fuel tanks is a concern, but it's also a reminder that pollution – most of which is invisible to us – now pervades the entire ocean, even at the point furthest from land.

THE SPACE INDUSTRY isn't alone in treating the high seas as a convenient waste disposal site, and the impact of splashdowns, at their current frequency, is minimal compared to other sources of ocean pollution. It is, however, emblematic of our 'out of sight, out of mind' attitude towards the high seas that has prevailed for hundreds, if not thousands, of years, thus allowing ocean pollution to become a growing threat – one that is now widespread, worsening and expanding in geographical area.

It's only been over the past several decades, however, that ocean dumping has become a serious issue. In the mid-1940s, a US-led campaign was launched to dispose of tens of thousands of tons of radioactive waste at sea. Between 1946 and 1970, 55,000 barrels of nuclear waste were laid down by the US at three sites in the Pacific Ocean. Other nations followed, including Belgium, France, Germany, Italy, Japan, South Korea, Netherlands, New Zealand, Sweden, Switzerland, Britain and most notably the USSR. Before long, the nuclear waste was added to, as nations started dumping material dredged from ports and rivers, waste from land-based mining, industrial chemicals and heavy metals, and ash from power stations at sea. In 1968, the US National

Academies of Science released a report that concluded over 100 million tons of waste had been dumped in the open ocean. The situation was out of control.

Over the half century that followed, a number of international rules came into effect to ban the disposal of noxious materials at sea. The first of these regulations targeted only 'high-level radioactive waste'. But, steadily, the list grew to include other hazardous materials, including low-level radioactive waste. Still, there was rule-flouting, especially by the Soviets, who continued disposing of vast amounts of extremely dangerous radioactive material, including nuclear reactors. Over the years, more international agreements have been signed, and more rules have been passed to prevent pollution at sea. The International Convention for the Prevention of Pollution from Ships (MARPOL), which came into force in 1973, for example, limits all sorts of noxious and harmful substances being disposed of from ships, including garbage, sewage and oil. The Stockholm Convention, which regulates the disposal of specific chemicals, lists twenty-eight banned or restricted pollutants, including dichlorodiphenyltrichloroethane (DDT), an insecticide used in agriculture that is deadly to marine life and seabirds. The London Protocol – a rule that is implemented by the UN's International Maritime Organization – has been in force since 2006, prohibiting the dumping of all waste in the ocean, save for some 'safe' materials, such as leftover materials from dredging. Alongside these laws, the Basel Convention controls the transport of hazardous waste such as corrosive and toxic material and infectious agents in international waters. On a national and local level, there are further regulations; the Clean Water Act in the US for, instance, sets a daily limit for industry on the amount and type of waste that can be released into the sea.

Despite the abundance of rules and regulations, pollution continues to enter the ocean, even far offshore, daily, without any consistent global limits. While deliberate ocean dumping has declined in recent decades, it remains a problem on the high seas: many nations don't abide by the rules, or lack the resources to dispose of waste properly; others never even agreed to the rules in the first place. Numerous industries get a free pass because they are important to national economies: the space industry, for one, and also the cruise liner industry, whose vessels can legally discharge around 1 billion tons of untreated sewage into the open ocean each year. What's more, even as dumping has declined, ocean pollution has grown and become more complex.

These days, 80 per cent of ocean pollution comes from land-based sources: it is runoff from farms – a toxic stew of nitrogen, phosphorus, fertiliser and sewage – mixed with industrial chemicals, pharmaceuticals and plastics. The remainder is mostly waste from shipping and offshore operations such as oil and gas platforms. Added to this are legacy pollutants, dumped at sea decades ago, but still surprisingly pervasive. Like zombies, they refuse to die. Mercury, for instance, a deadly poison that even in small doses can cause brain damage and heart disease, is one such legacy pollutant. Released from coal combustion and from gold mining, in the past 500 years the environmental load of mercury has increased 450 per cent above the natural background level. Currently, about 2,400 tons of mercury enter the environment each year, but this only accounts for about 17 per cent of the mercury found at the ocean surface; 50 per cent comes from pre-1950 emissions. Much like mercury and other persistent pollutants, DDT is still found lingering decades after its disposal; in recent years, it's been found clinging to ocean plastics, providing a new mode for its dispersal. In all likelihood, the pollutants that are now entering the ocean – chemicals in plastics, for instance – will be our mercury of the next century, infiltrating marine food webs, and our own food supplies, long after they first entered the ocean.

FOR MOST, THE ISSUE of ocean pollution intrudes on our daily lives only when it's visible, usually the result of a catastrophe such as the *Deepwater Horizon* disaster in the Gulf of Mexico in 2010, or the *Exxon Valdez* tanker rupture in 1989, both of which discharged huge volumes of heavy crude oil into the ocean. On seeing the immediate impacts, our collective concern is roused and, for a short while at least, we become aware of the fragility of our ocean and its marine life.

The *Torrey Canyon* shipping disaster was the first such incident to bring this message home to the general public. On 18 March 1967, a supertanker, flying the flag of Liberia and carrying 30 million gallons of crude oil, smashed onto rocks off the Cornish coast. In total, 110,000 tons of crude oil were released into the English Channel. The ship, named the *Torrey Canyon*, had been en route to Pembrokeshire from Kuwait, when the captain decided to take a shortcut; instead, he hit Pollard's Rock, a reef between Land's End and the Isles of Scilly, splitting open the ship's hull. By the time the *Torrey Canyon* was emptied of its load of crude oil, a 270-square-mile, foul-smelling slick of oil streaked across the ocean surface. Affecting hundreds of miles of coastline, the slick stretched to the Channel Islands, the coast of France and out towards the Atlantic Ocean. Before long, it had reached international waters.

The first major oil spill in European waters, it caused significant damage, killing at least 15,000 birds in the immediate aftermath, but taking a toll on marine life, from plankton to mammals, for at least fifteen years afterwards, and devastating local livelihoods. The impact of the event was so great that it ultimately changed the way people viewed, and governed, the oceans. The following year, in direct response to the *Torrey Canyon* disaster, two treaties were signed in Brussels, sponsored by a UN agency called the Intergovernmental Maritime Consultative Organization, which was the predecessor of today's International Maritime Organization, the global shipping regulator. Essentially, the new laws meant that nations could take measures – such as using chemical dispersants to break up an oil spill into smaller droplets – on the high seas, to reduce the chance of their own coastlines being polluted, and secondly, that all international ocean-going vessels would have to be insured by the owners, so that they could pay for any compensation and clean-up, in the event of a disaster.

The reality, however, is that even now, with all these rules in place, such acts of environmental negligence can be hard to prosecute, especially when they impact waters beyond national borders. A case in point is the *Sanchi* oil spill, to date the worst tanker spill of the twenty-first century. The incident happened on the afternoon of 6 January 2018, when the oil tanker *Sanchi*, flagged to Panama, owned by Iranians and crewed by Iranians and Bangladeshis, collided with a Chinese cargo vessel CF *Crystal* 185 miles off the coast of Shanghai, in the East China Sea. The tanker caught fire, exploded, and sank to the seabed, killing all thirty-two of its crew and spilling over 100,000 tons of condensate, a light, highly combustible crude oil. While most of the condensate burned off in the fire, some entered the ocean, creating a slick the size of Paris in international waters. The area where the tanker sank is a migratory route for three species of whale, and a place where numerous fish come to spawn. Unlike the *Deepwater Horizon* and *Torrey Canyon* spills, which involved heavy crude oil, the light combustible condensate carried by *Sanchi* had never before been released into the marine environment in such quantities – it was an entirely new threat with unknown consequences.

Instead of forming long-lasting slicks on the surface, condensate tends to burn off, releasing invisible chemicals into the water that can stay there for weeks or even months, spreading far beyond the accident site. In the case of *Sanchi*, scientists predicted that much of the toxic pollution would get caught up in powerful currents and make its way to Japan and out to the North Pacific Ocean. But because the accident happened on the high seas, involving a ship registered in Panama – a flag state with lax enforcement – that was owned by Iranians who were resident in Hong Kong, it's been hard to hold anyone accountable. In the aftermath, an investigation concluded that both *Sanchi* and the Chinese *CF Crystal* failed to keep look-out. While all parties involved agreed to take measures to avoid a similar incident in the future, no one to date has accepted liability for the lives lost or the environmental damage.

Accidents such as the *Torrey Canyon* and *Sanchi* spills are, by and large, caused by human error. A bigger environmental toll, however, comes from smaller incidents that are both frequent and deliberate, and the extent of which is only just now coming to light. For more than forty years, it's been illegal to dump oil into the ocean. That includes oily water, known as 'bilge water' containing other chemicals and detergents used in the running of a ship and which collects in the ship's hull, usually in quantities of several tons per day. By law, bilge water should be treated on board and disposed of on land in special facilities. But many ship operators flout this rule, instead dumping untreated bilge water straight into the sea. The practice is so common that this slow drip of oil into the ocean is causing chronic pollution, according to a recent analysis led by German news service Deutsche Welle.

Data provided by the environmental watchdog SkyTruth, suggests that, in 2021 alone, there were hundreds of individual instances of bilge-dumping across the global ocean. SkyTruth estimates that, every year on average, at least 52.8 million gallons of oily, toxic bilge water are being discharged illegally into the ocean. For comparison, the *Torrey Canyon* disaster released 25–36 million gallons, or roughly half to two-thirds as much. But the satellites used in SkyTruth's analysis cover only one-fifth of the world's ocean; vast swaths of the high seas are left out. Considering that 90 per cent of goods are transported across the ocean, by some of the largest machines on Earth – with engines as large as three buses stacked end to end – the actual footprint of these oil spills is likely to be much larger.

In shunning the rules, rogue ship operators have developed strategies to avoid detection, including the use of portable bilge pumps to discharge oily waste quickly and dumping only at night, in the open ocean or in rough seas, where the waste is harder to see. A 2016 study found that even small amounts of oil released at sea can have 'immediate adverse biological effects' on marine life, including reducing numbers of plankton. It can also affect feeding and reproduction in tiny sea creatures such as zooplankton, with consequences all the way through the food web. Legally, ships sailing international waters are answerable to their flag state and, to some extent, to the state where they berth their ship. But in international waters, transgressions are frequent – on the open ocean, this crime is hard to detect, enforcement is sporadic and even if a case was taken against a ship owner, the fines are pitifully low. Put simply, there's a cost incentive to illegally dumping at sea.

ENFORCING THE LAW offshore will only go so far, however, given that 80 per cent of ocean pollution starts out on land. Plastics; farming waste, in the form of excessive nutrients and fertilisers; and chemicals from industry get washed into coastal waters, and are eventually transported out to the open ocean by currents and by winds. Many of these pollutants, which routinely make their way to the high seas, have contributed to the creation of dead zones – vast areas that are so oxygen-depleted that they are devoid of life – throughout the seas. From 2008 to 2019, the number of dead zones in the ocean increased from 300 to 700, in part a response to nutrient runoff, but also due to climate change. About 30 per cent of our carbon emissions end up in the ocean, changing the temperature, and also the oxygen available to marine life.

As the newest pollutant entering the ocean in significant quantities, plastics mostly start out on land, with somewhere between 4.8 and 14 million tons making this journey each year, from the coast or down a river into the sea. First invented in New York in 1907, by a Belgian chemist named Leo Baekeland, the production of plastic didn't become mainstream until after the First World War; in 1950 global production was around 1.9 million tons per year; by 2018, that had risen to 465 million tons per year. The problem with plastic, of course, is not just we're producing so much of it, but that it takes centuries to degrade – up to 1,000 years for a single plastic bag. Of the roughly 8.8 billion tons of plastic that has been produced globally to date, already 7 billion tons, mostly created for single use, has become waste.

The first recorded land-based plastic waste found in the ocean was a bag, collected during a plankton survey off the west coast of Ireland in 1967. At the time, and for years after, plastic waste was a rare find on such surveys. But these days, plastic is so pervasive that it has been found throughout the global ocean – in the

Arctic, in the Southern Ocean, even at Point Nemo. Once plastic enters the ocean, it travels, and much of it becomes caught up in the ocean gyres that concentrate marine life, and now entrain our trash so successfully that they have become the ocean's great garbage patches. The most famous is the Great Pacific Garbage Patch located in the North Pacific Gyre between Hawaii and California; this area covers 620,000 square miles and holds an estimated 78,000 tons of plastic. If the outer, less concentrated centre of the patch is included, there is almost 100,000 tons.

The term 'garbage patch' conjures up images of large visible hard plastics – the ocean equivalent of a landfill. But most of the plastic waste at sea is microplastic, which is by definition anything smaller than 5 millimetres wide and much of it invisible to the human eye. Some of these microplastics are manufactured as such; one example is the tiny plastic beads commonly used in a face wash; other microplastics are 'nurdles' – lentil-shaped pellets that are used by industry to manufacture larger plastic items such as soda bottles, plastic bags, or milk cartons. Over time, these can be weathered down into microplastics. Gradually eroded by wind, waves and sunlight, these large plastics eventually become no more than tiny specks drifting on the currents, 99 per cent of which are below the surface. Ranging in size from an individual grain of rice to a submicroscopic virus, these particles are easily ingested by a whole range of creatures, from marine plankton to humans, and are now found throughout the world's ocean.

So, who is responsible for the plastic waste epidemic that's engulfing the sea? There are different answers to this question, depending on whether you place the burden of responsibility on producers of plastic, their users or waste managers. It also depends on where in the ocean you look for an answer. The statistic you'll hear most often is that 80 per cent of plastics started out on land – that includes soda bottles, plastic packing, tyres and so forth, with coastal developing economies such as Indonesia and Brazil as major contributors, owing to their laxer waste management regimes. What you'll hear less often is that just twenty companies are responsible for 55 per cent of the single-use plastic waste on Earth. Those twenty companies are petrochemical giants such as ExxonMobil (which tops the list) – the same companies that are polluting the ocean with CO_2 and whose oil is being shipped around in supertankers. So, most of the plastic bags and bottles that start out on land were produced, as a by-product of the petrochemical industry, by an oil major.

On the high seas, far from the coast, the origin of plastics becomes more complex. According to a recent study published in *Scientific Reports*, 86 per cent of floating plastic debris in the Great Pacific Garbage Patch comes from fishing vessels, and is mostly lost or discarded fishing gear. The study was led by The Ocean Cleanup, a Netherlands-based non-profit which has had success in reducing the size of the patch – in July 2022, they reported having removed over 220,000 pounds of plastic from the site (the dry weight of a space shuttle), a feat that repeated 1,000 times would eliminate the patch entirely. But they've also studied the contents of the patch – in 2019, during part of their clean-up operation, the team collected over 6,000 floating items totalling 1,206 pounds from one small area. By analysing the debris for logos and any other indicators of likely origin, they were able to trace 232 of the objects back to their home countries: topping the list were Japan; Taiwan; the US; South Korea; and the Chinese mainland, Hong Kong and Macau. This suggests that on the high seas, at least in the Great Pacific Garbage Patch, most floating plastics can be traced back to five industrialised fishing nations. But how to tell whether they were thrown overboard from vessels that departed these countries or simply made their way, from their shores, across the ocean?

To answer that question, the team used a computer model that simulates the likely dispersal pathway for debris carried on currents, and moved about by wind and waves. They ran the so-called 'Lagrangian dispersion model' for different scenarios, some in which the plastic originated from land and others in which it was disposed of by fishers at sea. From this exercise, they concluded that most plastics had come from vessels already far from shore. Because scientific observations of debris at sea are few, these sorts of dispersion models can help fill the gaps and test various ideas about how plastics move around in the ocean. Scientists have used them to retrace the likely journeys, and origin, of all sorts of items lost at sea – including a flaperon, or hinged flap, that broke off the plane that was Malaysia Airlines Flight 370 and turned up, sixteen months after the plane's disappearance, on Réunion, an island in the western Indian Ocean.

CONVERSELY, SCIENTISTS ALSO use adrift objects to learn more about the inner workings of the ocean at different scales – how currents work to move heat, salt and, increasingly, our waste around the planet. The pioneer of this research is an American oceanographer named Curtis Ebbesmeyer, who studies the movement of flotsam or floating objects, most of them items lost at sea from shipping containers, to track ocean currents. Ebbesmeyer first launched the discipline, which he calls 'flotsametrics', back in 1991, in response to a question from his mother during a routine visit to her house in Seattle. It was June and holidaymakers crowding to beaches along the Pacific coast of Vancouver, Washington and Oregon were struck by an unusual phenomenon: hundreds of Nike trainers, seemingly brand new, were washing up on shore, creating a frenzy among locals to assemble pairs of the expensive shoes. With no convincing explanation as to how the Nikes had arrived here, Ebbesmeyer, an oceanographer, was challenged by his mother to find out.

Ebbesmeyer soon traced their origin to a shipping incident the year before, when a freighter, the *Hansa Carrier*, had met a storm midway between Seoul and Seattle and lost several shipping containers, twenty-one in total, to the frigid waters of the North Pacific Ocean. Five of the lost containers were filled with Nike brand trainers; one container sank to the seabed and four opened on impact, releasing their contents to the great big blue. That is how 61,280 shoes each began their solo journeys at sea. For Ebbesmeyer, whose interest was in understanding ocean currents, this was a windfall: he soon realised that each Nike shoe has a distinct serial number, making them eminently traceable. Nike's transportation department provided Ebbesmeyer with all of the information he needed from the spill: the latitude and longitude, the container load plan, and each container's contents.

Armed with this intelligence, Ebbesmeyer knew the point of release of the 'drifters'. Now all he needed to find out was where the currents were taking them. For this, he relied on a network of volunteer beach-combers, who offered to keep an eye out for the trainers while scouring the world's shores, and report back. Ebbesmeyer, in turn, then entered the data into a computer model called the Ocean Surface Current Simulator (OSCURS), which had been developed by a college peer of his named Jim Ingraham. He tested the model by giving Ingraham only the data on the point of the shoe spill, and asked if he could calculate when and where the Nikes would wash up. Within an hour, Ingraham had guessed the earliest arrival spots for the trainers – November and December 1990 on the Washington coast and January and February 1991 on Vancouver Island.

Before long, another opportunity presented itself. On 10 January 1992, a violent storm hit the Pacific Ocean and a freighter travelling from China to America lost several shipping containers in the squall. One of these containers was filled with 28,000 plastic bath toy sets. Each set contained four figures – a yellow duck, a green frog, a red beaver and a blue turtle – mounted on hard board, and plastic wrapped. Over time the wrapping and cardboard were worn down by ocean waves, setting most of the 'Friendly Floatees' free, bobbing to the ocean surface to begin their transoceanic journeys.

Almost a year later, the first Floatees had made their way across 2,000 miles of open ocean to the shores of Alaska. Over the next year, several hundred more followed, these ones travelling to the eastern coast of the Gulf of Alaska. Again, OSCURS foresaw their likely journeys, and predicted their arrival in Washington state a couple of years after Alaska. The model also predicted that, within several years, ducks, beavers, frogs and turtles would be crossing the Arctic Ocean, making their way into the Atlantic. Sure enough, in 2007, the weather-worn bleached bath toys, still recognisable from their branding, turned up on the southwestern shores of the UK.

Since then, other objects have joined the flotsametrics study, among them a consignment of 4,756,940 LEGO pieces lost from a container ship that ran into trouble off Land's End, UK, in 1997. Among the small coloured plastic pieces were 33,427 black dragons and, amazingly, thousands of miniature ocean-themed LEGOs, including scuba tanks, flippers, octopuses, sharks and assorted shapes used for underwater 'scenery' including Little Ugly Rock Pieces and Big Ugly Rock Pieces. They are still washing up on Cornish beaches.

Objects have, of course, been falling into the sea and washing up on the shores since humans first braved the offshore world, and for a lot longer if you take account of driftwood, volcanic pumice and other natural materials that make their way out to sea. The difference now is that much of what's entering the ocean is so enduring, being transported around the globe in many cases for centuries, or longer. While all of this flotsam has helped oceanographers such as Ebbesmeyer track currents as they move around, it's also helping scientists to understand how the ocean's physical machinery is transporting plastics and other pollutants, information that is sorely needed for assessing the risks to marine life.

At its largest scale, this machinery takes the guise of a global circulatory system that is powered by regional variations in ocean salinity and heat. Known as the thermohaline circulation, or more commonly as the 'global conveyor belt', this series of ocean currents is how heat is transported from the tropics to high latitudes, warming northerly countries and maintaining their temperate climates. But this same system, which acts as the lifeblood of the ocean, transporting nutrients, sediments, and plankton to new regions across the globe, is now also responsible for transporting pollution, including plastics, to the far corners of the ocean.

For at least 3 million years, this conveyor of heat and salt has operated with varying intensities, at times switching off entirely. In its current state (weakening, though still conducive to the climate as we know it) it begins in the far north of our planet. Confronted with near-freezing Arctic temperatures, seawater here becomes chilled. Topped with sea ice – which sheds salt into the water, making it denser – it sinks. Laden down, this cold, heavy water eventually reaches the ocean floor off the coast of Greenland, forming deep forceful channels that carry as much water as eighty Amazon Rivers. This heaving body of water begins a slow journey southward, past entire continents, across the equator, through the South Atlantic and into Antarctica's Southern Ocean. On reaching Antarctica's Weddell Sea, it gathers strength, becoming ten times larger, more powerful, until eventually it is carrying, effortlessly, as much as 48 billion gallons of water per second, more than 100 times the flow of all the rivers on Earth. From there it spreads out, stretching from the ocean depths right up to the sea surface, and envelops the entire Antarctic continent like a protective impenetrable shield. On its continent-facing side, the conveyor belt is icy cold. Here, the belt is 'recharged', picking up cold, dense salty water that allows it to sink – as it did, before, at the other end of the world, in the Arctic – and gather momentum for the return leg of its journey.

Some of this heavy fluid mass sinks to the bottom of the seafloor and from there flows northward deep in the ocean as 'Antarctic Bottom Water'; staying below 9,800 feet, it spreads out across the global ocean as far north as the equator. On its outward-facing side, the ocean conveyor belt steadily warms, however, and pushes towards the Indian, Pacific and Atlantic Oceans to begin the northward, return leg of its journey. Eventually, this water, having warmed and freshened on its homeward stretch, reaches the Arctic and the cycle begins again.

In 2017, an international study found that plastic debris is now hitching a ride on this global conveyor belt as it travels northward to the Arctic, with the result that microplastic pollution is now accumulating in the Greenland and Barents Seas, which act as a dead-end for the conveyor belt on its return leg. The microplastics that end up here have, according to this one study, been in the ocean for decades. Separate research has shown that microplastics have been present in Antarctic waters too, since at least 2009, presumably carried there by the ocean conveyor belt on its southbound voyage.

But deep ocean convection is only one part of the ocean's intricate physical machinery; operating on a smaller scale is a network of regional, speedier circulatory systems known as gyres that stretch thousands of miles across. One of these is the North Pacific Gyre; the site of the Great Pacific Garbage Patch. By their very nature, gyres entrain marine life, and therefore also debris – similar gyres and garbage patches exist in the South Pacific, the Indian Ocean and the Atlantic. Each is a collection of currents that spins in a circular motion within an ocean region – such as the North Atlantic or the South Pacific – propelled by winds and by the Earth's rotation. On their western flanks, gyres give rise to boundary currents – powerful, steady streams of water that transport heat, and pollutants, from the equator to the poles. These western boundary currents include the Gulf Stream of the Atlantic, the Kuroshio Current in the North Pacific and the Somali and Agulhas Currents of the Indian Ocean.

On their eastern side, gyres spin out broader, deeper currents that, in opposition to their western counterparts, move cold water, formed deep at the North and South Poles, towards the equator. Running parallel to the coast, these eastern boundary currents are mostly driven by local coastal winds, but are also pushed offshore by the Earth's rotation. Rich in nutrients, these currents comprise some of the most productive waters on Earth, and include the Humboldt Current, off the coast of Peru, and the Benguela Current, along Africa's west coast. But within any ocean basin, there are more transient forces at play than deep-water convection and gyres. Eddies are smaller whorls that typically break off from currents or form independently. Spanning around 50 miles across, they persist anywhere from ten days to a year and, at any given time, can number several thousand in the open ocean. Once born, they can travel for months and cross hundreds or even thousands of miles, bringing nutrient-rich water from the depths to the surface, where they fuel conditions for life in the open expanses of the high seas.

This system of transport can carry our waste surprisingly far: the conveyor belt can literally transport pollution from one pole to another, boundary currents can bring pollutants from the poles back to the equator, gyres can entrain trash within ocean basins, while individual eddies can carry a piece of plastic across literally an entire ocean basin. In the aftermath of the Tōhoku-Oki earthquake and tsunami that struck the eastern seaboard of Japan in 2011, as much as 22 million tons of marine debris was ejected into the North Pacific Ocean. Heavy items sank close to shore, but several million tons of buoyant miscellany – derelict vessels, a local village sign, a motorbike in its container, and smaller items such as bottle caps – made their way out to sea.

Over the following years, much of that landed on the shores of the North American Pacific Coast and on the Hawaiian islands, having travelled 3,700 miles across the Pacific Ocean. An unexpected consequence of this mass of debris was that the individual pieces doubled up as makeshift rafts for tiny ocean creatures that ordinarily would cling to a rock or the frond of a seaweed on a beach. Species have always clung to ocean debris, whether driftwood or seaweed, and there have always existed creatures that use unusual means to navigate the high seas. The violet sea snail (*Janthina janthina*), also known as the bubble raft snail, for instance, uses its muscular foot to create a floating raft of bubbles from mucus, from which it hangs suspended, upside-down, while drifting across the ocean. *Velella velella*, the by-the-wind sailor – a close relative of the jellyfish – uses a tiny sail that pushes it far offshore on long-distance voyages. This time, however, hundreds of individual marine species clung to the debris and found themselves months or even years later washed up on some distant shore, some as far adrift as Hawaii. Not only did these Japanese coastal species survive and grow at sea, but they even reproduced on tsunami debris in the open ocean, forming their own ocean-rafting community.

Some thriving clingers-on eventually found their way to the Great Pacific Garbage Patch, prompting scientists to take a closer look at whether floating plastic debris is, more generally, emerging as a new habitat on the high seas. That was how, in 2021, scientists from the Smithsonian Environmental Research Center came to describe, for the first time, an entirely new type of ocean community, which they named the neopelagic – 'neo' for new and 'pelagic' for open ocean. A mish-mash of coastal and oceanic species, this new community thrives not just on plastics that began their journey on land, such as the tsunami debris, but also on plastics discarded on the high seas, such as derelict fishing gear. The deluge of plastics that has made its way to the high seas is now transforming the communities that live, and navigate, the global ocean.

As a threat to the high seas, unintentional pollution can be tough to tackle. It's a problem that starts far away and by the time its impact is seen, the original transgression has long passed, the scene of the crime often unknown. What we are left with is a garbage patch larger than a US state, a growing dead zone, or an alien community washed up on a foreign shore – and the realisation that the high seas, despite our imagined boundaries, are connected to the rest of our planet. As far as the industries responsible are concerned, they see their contribution as being comparatively minimal – the shipping industry, for instance, points to the fact that most vessels abide by the rules, and that container loss is a rarity relative to the number of containers shipped. Similarly, the aerospace industry regards splashdowns as largely harmless compared with the pollution caused by other industries, such as shipping. And despite rules that have now been in place for decades, we still treat the high seas as a convenient dumping ground for our vilest junk. As evidence, take Japan's release, starting in August 2023, of over 1 million tons of contaminated waste water from the Fukushima power plant into the Pacific Ocean. The justification is that the contaminated water will be treated to a level deemed safe, but neighbouring countries and environmental groups have, predictably, expressed consternation.

In support of the Japanese plan, director general of the International Atomic Energy Agency, Rafael Mariano Grossi, said in 2021, 'Releasing into the ocean is done elsewhere. It's not something new.' In other words, we've already set a precedent for hiding our most noxious waste in the ocean – what's to stop a little more? Much as the ocean was once thought too boundless to exploit to exhaustion, it was also thought capable of endlessly absorbing our trash with little impact. We're now faced with a new realisation, that our vast, deep ocean is incredibly fragile and its greatest threat is us.

10

THE COLD RUSH

A FEW YEARS AGO, Emirati businessman and self-described entrepreneur Abdulla Alshehi proposed an outlandish scheme to solve his nation's growing water crisis: he would tug an iceberg weighing 44 million tons and measuring 1.2 miles long, 1,650 feet wide and 650 feet deep – around the size of a small island – from Antarctica's Southern Ocean to the Fujairah coast in the Arabian Gulf, a trip of around 8,700 miles. If successful, this would offer the United Arab Emirates potable water for a year – a valuable commodity for a region that is becoming increasingly parched. Alshehi's berg would be selected by satellite and hauled to a position around 2 miles offshore, from where the colossal ice block would be chipped and crushed into smaller pieces until its bulking 110-million-ton structure was thawed, bit by bit, and added to the national water supply. Ideally flat and tabular in shape, the berg, though enormous, would, in theory, be relatively easy and safe to transport. The trickiest part of the operation – hauling the colossal structure across the high seas – would be aided by a metal belt, a device that Alshehi, a trained engineer, has designed and successfully patented, but has yet to try out in the real world.

If this sounds far-fetched, it's because the United Arab Emirates, Alshehi's home country, and the entire Middle East, is increasingly desperate: in the past fifty years, drought has sucked half the water supply out of several major rivers, and temperatures on land can now exceed 122°F. In consequence, the UAE now consumes about 15 per cent of the world's desalinated water, a costly solution that comes with its own environmental problems. Meanwhile, around 75 per cent of the freshwater on Earth is locked up in icebergs, and 90 per cent of that ice is in Antarctica. In an average year, Antarctica sheds around 140,000 icebergs, and Greenland sheds around 20,000.

Cumulatively, these icebergs add around 326 quadrillion gallons of freshwater to the ocean each year. By one calculation, just 5 quadrillion gallons of iceberg water, harvested annually, could quench the thirst of the approximately 5 billion people who live in water-stressed regions, where demand outstrips supply for at least one month of every year. In that context, harvesting water from icebergs may seem a sensible proposition – it's a resource that we either use or lose, after all. 'Many icebergs are disintegrating from Antarctica', Alshehi said recently. 'Once they do, they float in the ocean and melt, wasting billions of gallons of fresh water. So we thought why not utilise it?'

But as reasonable as this may sound, little is known about the consequences of tugging a several-thousand-ton melting ice block across the high seas and parking it offshore, in some of the warmest waters on Earth. For a start, it would likely require several tens of thousands of tons of fuel for a 100-day journey, according to marine services company Atlantic Towing. Once the iceberg is parked in temperate or warm waters – such as offshore of the Fujairah coast – it might change the region's microclimate, by cooling the surrounding waters and atmosphere. Icebergs that make their way to warm climes will invariably freshen the surface of the ocean; because freshwater has lower density, this forms a surface water cap and lessens ocean mixing, with unknown impacts for the creatures that live below. The individual effect may be negligible, but done with any regularity, iceberg harvesting may have a discernible impact. With little in the way of evidence, however, scientists can only speculate on the impact of having misplaced icebergs dotted around the ocean.

There's also the issue of cost: Alshehi's scheme has an estimated price tag of around $200 million, a price point that Stephen Bruneau, professor of coastal and ocean engineering at Memorial University in St John's, Canada, described as being 'so utterly, utterly uneconomical as to be ridiculous'. Others see it as 'an absurd waste of money'. Right now, however, the potential downsides are beside the point: legally, icebergs are an unregulated resource of the high seas, free to be taken, by anyone who has resources and technology to organise such a scheme. In that sense, they are much the same as mesopelagic fish harvested from the ocean's twilight zone or the squid fisheries being targeted in regions such as the western Indian Ocean.

Strictly speaking, anyone towing an iceberg across an ocean should gauge the potential damage, but no one is enforcing that rule on the high seas. As such, the Emirates claim to have looked at the likely environmental fallout, but they haven't been required to make that information publicly available. Nevertheless, if Alshehi succeeds at this iceberg-towing scheme, he will be the first and others will surely follow. As the world warms – and the polar regions thaw – one possibility is that there will be more and more ice to be captured, towed and harvested. According to some experts, such as Matthew H. Birkhold – an environmental humanities professor at Ohio State University – a race to harvest the ocean's icebergs may soon be upon us. If this still sounds far-fetched, consider the fact that icebergs are already being harvested on a smaller scale.

The first time Ed Kean went searching for bergs he was age six, with his father. Back then, they were just looking. He remembers being awestruck by the huge cathedrals of ice that found their way south, on ocean currents, to the shores near their home in Newfoundland. Kean's people were fishers – his father, but also his grandfather and several uncles. At first, he followed suit, becoming a fifth-generation sea captain, the proud skipper of a small boat called *Green Waters*.

Today, his quarry isn't fish, but the icebergs that first captivated him all those years ago. 'Treasures of the sea' is how he describes them. He began harvesting and selling icebergs in 1972, first to local businesses – breweries and distilleries that prized the pure water for making, and marketing, their beers, gins and vodkas. More recently his customers have grown to include local cosmetics and wine producers. Now in his sixties, Ed Kean, the 'Iceberg Cowboy', has launched his own brand, a luxury water that he hopes to export to high-end markets in countries such as China and Dubai. 'It's definitely the best-tasting water you're ever going to drink', he told me. 'This is water that was frozen anywhere from 15,000 to 25,000 years ago. That means the Industrial Revolution hadn't occurred, and there were no pollutants in the snow.' Most of the bergs that Kean harvests start their lives on Greenland's west coast, as part of the Jakobshavn glacier, a mammoth wall of ice some 40 miles long and about a mile thick. From there, they set out on a journey of 2,000 miles, battling the waves and warming waters as they head south through 'Iceberg Alley', a treacherous passageway that runs from Greenland to the southern coast of Newfoundland and Labrador.

For Kean, the season is short and the work dangerous, squeezed into six backbreaking weeks, in June and July, each year. Once the bergs arrive, those that aren't captured melt into the ocean within a couple of weeks. It's a race to retrieve as much ice as possible. It's not easy catching icebergs on a rocking boat. 'Nine-tenths of an iceberg lie under the water and you don't want to flip it', Kean said. Though it's not uncommon to have million-ton bergs in Iceberg Alley, he targets the smaller ones, which weigh in at around 5,500–6,600 tons. They go by different names, depending on their size. Small bergs are generally called 'bergy bits'; one less than 6.5 feet across is a 'growler'. Kean nets the smallest ones like giant immobile fish, first lassoing them with the net and then hauling them onto the ship's deck.

Sometimes he approaches the largest ones with an air rifle, blasting them in the hope that a great chunk will break free – an easy win. But mostly, he uses an excavator, a machine that breaks off great hunks of ice. Icebergs can be unpredictable. 'They roll over, they roll out and they roll in and they implode and explode.' The greatest threat is a mini tsunami, which can topple the boat. 'We've had a few close calls', he told me, relaying an incident where he was on top of the excavator trying to grab a slab of ice when the berg began to roll. 'There was a huge wave and everything went helter-skelter for a while', he recalled. Once the ice is on the boat, it's broken up into 265-gallon chunks, and shipped back to a shore-based processing facility, where any natural impurities – mostly volcanic ash or microbes – are removed.

Each spring, as many as several hundred, even a thousand, bergs can arrive at the shores off Newfoundland. Kean harvests just a fraction. In a good year, his crew collect around 210,000 gallons of water, which they'll sell on for a dollar a litre. Over his years in business, Kean and his crew have extracted about

4 million gallons of water. 'The bit of ice that we take now is not that much; it equals a million litres at most', he said. The Jakobshavn glacier alone calves more than 10 cubic miles of icebergs into the ocean each year, a quantity that could, if needed, satisfy the annual potable waters needs of the entire US population. Considered this way, it is hard to see iceberg harvesting as being especially harmful. But with waters warming, this nascent industry is attracting other opportunists, and some with much larger ambitions than Ed Kean.

THE IDEA OF towing icebergs across the high seas was first proposed in the early 1800s as a way of 'equalising the temperature of the Earth'. It wasn't until the mid 1800s, however, that anyone attempted to move an iceberg any sizeable distance. The first to do so were the Chileans who captured small bergs south of Chile, fitted them with sails to make use of the prevailing winds and towed them to Valparaíso, where there was a booming brewing industry. Around the same time, bergs were being floated from Alaska to California. Within a few years, the idea of floating bergs as far as India was being discussed with great fervour in magazines such as *Scientific American*. Nothing came of these more ambitious plans, however, and almost a century passed before the idea took hold again; in 1956, an oceanographer named John Isaacs at the Scripps Institution of Oceanography in California suggested towing an 8.8-billion-ton iceberg from the Southern Ocean to San Clemente Island – 75 miles west of San Diego – in response to a local drought. The scheme would take 200 days for a berg measuring 20 miles long, 3,000 feet wide and 10,000 feet deep, and would provide the city with an immediate source of freshwater. As Isaacs saw it, the energy involved in the tow would have been comparatively little relative to that needed to desalinate an equivalent amount of water, which was an alternative approach emerging at the time. Isaacs' proposed scheme gave renewed life to the notion that icebergs – positioned in the right places – could solve the world's water shortages.

Then in the 1970s, Mohammed bin Faisal Al Saud, a Saudi prince, decided to take the idea a step further. He commissioned French engineer Georges Mougin to assess the feasibility of a tow from the Southern Ocean to the shores of Saudi. Presenting his findings at a niche scientific congress in Iowa in 1977, Mougin – also a renowned glaciologist – attracted the attention of some serious polar scientists: a Norwegian named Dr Olav Orheim, and Professor Peter Wadhams, who would later become the director of the Scott Polar Research Institute at the University of Cambridge. But despite the growing interest, it soon became clear that the technology didn't exist to tow a big berg from Antarctica to the Arabian Gulf. The idea languished after that, until 2010 when Mougin teamed with a software company named Dassault Systèmes to resurrect his plan, breathing new life into it with a computer-generated simulation of his proposed tow. Mougin wanted to test the feasibility of hauling a 7.7-million-ton iceberg wrapped in 'insulated fabric mesh' from Newfoundland to the Canary Islands; the venture with Dassault Systèmes used satellite data and 3D modelling to simulate the journey under reasonably realistic conditions, including, for instance, one fairly strong mid-Atlantic storm.

Their calculations showed that the tow could be done within 141 days, would require around 4,400 tons of fuel and would cost $10 million. It was this analysis that piqued the interest of Alshehi in the UAE. In 2015, the Emirati published a book called *Filling the Empty Quarter: Declaring a Green Jihad on the Desert*, in which he documented his vision of transforming the Arabian Desert into a verdant green pasture, with free-flowing rivers and lakes throughout the UAE, and a 'Great Green Wall' – the first of its kind in the world – aided by freshwater imports. Needing a reliable and sizeable source of water to realise this plan, Alshehi seized upon the idea of iceberg harvesting. Compared to the infrastructure needed to pipe this water to the UAE, or to build further desalination plants, tugging an iceberg across the high seas seemed a comparatively straightforward proposition. In Alshehi's scheme, the iceberg would also be inhabited by populations of both polar bears and penguins, left to roam free along the shores of the UAE.

The other contender in the race to tow the world's first mega berg is Nick Sloane, a marine salvager who has wrestled with the wrecks of tugboats, oil rigs and planes. Most famously, the sixty-year-old South African oversaw the 2014 re-floating of the huge sunken Italian cruise ship *Costa Concordia* off the coast of Isola del Giglio in Tuscany. In 2018, amid Cape Town's water crisis, Sloane offered up his services for the purpose of quenching the city's thirst. At the time Cape Town, which had been in drought since 2015, was due to reach Day Zero, the point at which the city's taps would be turned off, forcing locals to queue for water at guarded standpoints. Sloane's proposal was to tow a giant iceberg from Gough Island in the South Atlantic Ocean just north of Antarctica to the city and melt it down as drinking water. The berg Sloane was targeting was 95–110 million tons, twice the size of the UAE mission and ten times larger than the Tenerife project. By his calculations, it would have cost $130 million and could have provided 40 million gallons of water each day for a year, meeting 30 per cent of the city's needs.

In January 2023, Alan Condron – a geophysicist based at the Woods Hole Oceanographic Institution in Massachusetts – published his own evaluation of the feasibility of both the Alshehi and the Sloane iceberg-towing schemes, based on calculations using an independent numerical model named MITberg and specifically designed to study the feasibility of iceberg towing. The MITberg model factors in ice loss due to exposure to sunlight and wave induced erosion, for instance. It also simulates realistic volumes of ice calved from the Antarctic ice sheet, producing an iceberg drift pattern and a total volume of ice that closely matches what's currently observed in the Southern Ocean. Condron's conclusions, published in the research journal *Scientific Reports*, are that, while feasible, tugging a berg large enough to supply drinking water for a year to Cape Town, for instance, would require three vessels and, on arrival, would need to be harvested or stored almost immediately. Within forty-four days, it would have melted entirely. Alshehi's scheme, on the other hand, would require ten to twenty tugboats to pull the iceberg across the high seas but, if harvested quickly enough, could supply the region with enough freshwater for six to nine months.

Despite all the noise, no one has actually hauled an iceberg across the ocean. On the high seas, including the Southern Ocean, icebergs are an unregulated resource, but once they enter national waters – either of their own accord, by hitching a ride on ocean currents, or having been lassoed and hauled there, they can be claimed – just like fish stocks. Right now, so little iceberg water is harvested each year that it doesn't yet matter who owns it. Yet that could soon change. In recent years, the UN has started describing icebergs as an 'unconventional water resource', perhaps in recognition of their growing significance in a water-stressed world. In 2021, the World Economic Forum listed iceberg harvesting among its top five alternative methods for countering water stress. 'I see icebergs as an interesting legal quandary. I think nations will start to position themselves as owning them in different ways', said Anita Lam, a Canadian social scientist who researches the ecological costs of consuming icebergs. One successful tow across an ocean could spur a long-anticipated industry, with bergs being moved around to meet the needs of increasingly parched communities under the effects of climate change. Those nations that harvest icebergs already, namely Newfoundland and Greenland, are now starting to consider how this industry should be governed. In recent years, the Newfoundland and Labrador Water Resource Management Division have also introduced some best practice guidelines. One rule, for example, mandates that harvesters should mark the bergs they plan to collect in a season and should only take one at a time. For the time being, harvesting happens at such a low level that this isn't enforced. Operators such as Ed Kean, for instance, must harvest their bergs out of sight of 'known locations frequented by tourists' and limit their take to 265,000 gallons per season. In time, the local government says it may even cap the number of operators. Greenland, on the other hand, has taken a different approach – rather than limiting licences, the government is interested in maximising the revenue from this emerging industry and requires operators to have a minimum annual harvest.

Icebergs that are free-floating on the high seas, beyond national jurisdiction, are a different matter. Who will get to harvest these common resources? Which communities should benefit? According to two environmental lawyers, Ana Costov and Jessica Appelmann from the University of Groningen, we've reached a point in time where we need to reopen that discussion. The situation, as they see it, is that we haven't decided who should benefit from this precious, untapped, water resource: is it everyone's? Should it be subject to some benefit-sharing rule? Or does a finders keepers rule apply? The global water crisis is now worsening to the extent that by 2025, as much as two-thirds of the global population may be facing water shortages. Meanwhile, enshrined in the UN Sustainable Development Goals is an ambition that everyone on this planet should have access to water for drinking and sanitation by 2030 – only a few years from now. At present, the only harvesters of iceberg water – such as Ed Kean – are selling the commodity on to luxury brands. They are harvesting a tiny fraction of what's out there, but as water scarcity becomes more severe, and the technology to move and exploit icebergs improves, even this level of extraction could become contentious. According to Matthew Birkhold, an environmental humanities expert, such high-end brands could ultimately force regulators to enact new laws dictating who owns this free, but increasingly commoditised, resource.

The current appeal of iceberg water is, in part, that it's freely available in huge quantities. But it's more nuanced than that. According to Anita Lam, the Canadian social scientist, iceberg water represents purity in a world that is increasingly polluted and altered by humans. In recent years, a slew of iceberg water brands has come online, aimed at those with an interest in savouring unusual, and untainted, water. They have been a big hit in the growing wellness movement. 'This is a natural resource that is being commoditised for different markets. Mostly, it's being sold as a high-end product', said Lam. Increasingly, the market is becoming flooded with brands, such as Ed Kean's own product. The best-known is Svalbarði, a brand launched in 2015 by Jamal Qureshi, a former Wall Street analyst. Retailing at $166 per litre, it's among the world's priciest waters. Berg Water, the Canadian equivalent of Svalbarði, retails on Amazon for $492 for a twelve-pack of 750-millilitre bottles. Harvested from 15,000-year-old North Atlantic icebergs, it is described as having a 'light and clean taste'. 'It's about access to an imagined kind of pure water', said Lam. The iceberg that sank the *Titanic* entered this world, calved from an Arctic glacier, around the time of Tutankhamun's birth. Icebergs are old and pure, in a pre-Anthropocene sort of way.

Icebergs have, of course long captured the human imagination for their beauty – variously described as 'masses of beautiful alabaster' and as 'one big white mystery' – and for their menace. In Samuel Taylor Coleridge's 1798 *The Rime of the Ancient Mariner*, he described the terrifying sound of a lurching iceberg as 'it crashed and growled, and roared and howled, like noises in a swound!' Since the sinking of the *Titanic*, in Iceberg Alley in 1912, icebergs have increasingly been vilified. E.J. Pratt, a leading poet at the time of the *Titanic* disaster, portrayed the offending berg as a zombie-like creature that could 'lurch and shamble like a plantigrade'.

These days though, icebergs are social media stars. In a climate-altered world, they have become a symbol of vulnerability. Formed by calving events, when a mass of ice breaks off from a glacier, icebergs now represent a weakened planet, which shunts these icy behemoths into the ocean with dismal regularity. Far from static, they morph and shapeshift as their bodies become fragmented over time. The uniquely shaped bergs that Ed Kean tackles in Iceberg Alley are described as 'non-tabular', meaning that they come in a variety of astounding architectural forms, all spectacular. Eager tourists capture them on camera and upload them to IcebergFinder, a website maintained by the Newfoundland and Labrador Department of Tourism. The most attention-grabbing bergs make their way to social media sites such as Twitter. The penis-berg, a phallic-shaped mass known on social media as #hammerofthegods, stole the limelight in 2017, for instance, but was replaced by the #beautyberg in 2018.

Tabular bergs, on the other hand, are flat and large; these can be the size of a small country. Free-roaming masses, they can weigh more than 20 billion tons and reach the height of a skyscraper. In July 2017, for example, a tabular berg measuring 2,300 square miles – 100 times the size of Manhattan – broke off from Antarctica's Larsen-C ice shelf, and began to drift northward on the current. At the time it calved, A-68, as it was named, was the world's biggest berg, weighing roughly 1 trillion tons. Geraint Tarling, a biological oceanographer at the British Antarctic Survey, was one of a small group of scientists who monitored the berg over its 3.5-year lifespan. Specifically, the scientists wanted to assess how the ice giant affected the environment – information that could indicate the impact of long-distance tows.

Tarling told me that A-68 hitched a ride on the Antarctic circumpolar current and made its way through the Weddell Sea before it became lodged off the coast of South Georgia. Parked offshore, it started to melt. 'It's an absolutely massive iceberg. Hundreds of miles long. The size of a small country', said Tarling, explaining that what was unusual was the fact that it stayed intact for so long.

At its peak, A-68 was dumping 1.6 billion tons of freshwater into the ocean every day, 150 times the daily consumption of the UK. 'The melting of an iceberg brings a lot of new things into a marine system', he said. 'It can actually affect the way that the oceans mix, the temperature of the waters and even what organisms exist in that environment.' But it's not all bad, added Tarling. 'It also brings in lots of nutrients that weren't there before and that's a good thing, because it increases the productivity of the environment; it allows more things to grow, to reach adulthood and to reproduce.' On the high seas, and especially in warm waters – which tend to be low in nutrients – this could have benefits. But there are risks. One is that water melting off an iceberg would be a lot like the ballast water that is used to stabilise a ship and that, before the introduction of regulations in 2004, was typically released from a ship into the ocean, bringing new species with it. 'An iceberg has its own particular types of flora and fauna, because of the fact that they live within the ice themselves', said Tarling. 'These flora and fauna would be offloaded into the water.' There's also the fact that the freshwater would form a top layer on the ocean, which changes everything that lies beneath. 'Deeper organisms rely on the surface layers for their food, because that's where phytoplankton live, that's where photosynthesis takes place. And if that's not happening, if photosynthetic organisms cannot actually thrive in that surface layer, that would trickle down to the rest of the food web.'

FOR THE TIME BEING, it's unknown whether a long-distance iceberg haul is even feasible. Proponents such as glaciologist Peter Wadhams recognise the limitations. Wadhams would like to see a berg towed from the near Arctic and brought to Tenerife, in a real-world version of Georges Mougin's computer-simulated scheme. A feasible concept is to capture one of the larger bergs from Iceberg Alley, just off the shores of Newfoundland. 'You'd be crossing the Atlantic and running with the Gulf Stream', said Wadhams, who is not only confident that this plan would work, but, given the global water crisis, he thinks we are compelled to give it a go. 'It's a horrible crime that this is being wasted. An iceberg melts anyhow . . . it's not like if you didn't tow it, it could be preserved somehow in a museum. It's melting all the time!' he said. But towing a berg as far as Saudi feels like wishful thinking to Wadhams. 'If you take a berg from a frozen sea, you need to take it to a cold sea, or at least not to a warm one. The water temperature is the killer.'

Others share his concerns. Stephen Bruneau, a professor of coastal and ocean engineering, admits that he would be quite taken with the idea of relocating bergs to drought-stricken regions, if he didn't have so much personal experience of the challenges involved. In the early 70s, Bruneau's father launched the world's first serious iceberg-towing expedition. A civil engineer by training, Bruneau – like his father before him – now teaches at Memorial University in St John's, Newfoundland's capital. Bruneau came to live in St John's as a young boy when his father became Dean of Engineering at the university.

At the time, Canada was keen to establish an offshore oil and gas industry but, unlike the North Sea, Canadian waters had a unique threat unfamiliar to the Brits and the Dutch – menacing towers of ice. So, Bruneau's father started to develop systems to detect and track the icebergs. Like weather forecasting, this involved using numerical prediction models to see the icebergs' likely routes and to gauge whether they could be 'managed'. The routes were often hard to predict, Bruneau explained, influenced as they are by wind and currents, and buffered by sea ice. 'Even today with supercomputers, we can't figure out exactly where they will go', he told me. His father, among others, discovered that it was possible to redirect an iceberg, but only by degrees; it wasn't possible to control them entirely. 'They proved that you could lasso an iceberg, you could deflect it slightly and change its trajectory, but you're not getting it out of Dodge', he explained. 'It was far from perfect, far from reliable ... and you could have a lot of sea ice at the same time, which is a different proposition altogether. That might make it so that you can't even get a tow line around the ice.' This meant that the oil and gas industry had to design rigs and ships that could be disconnected and moved quickly if a berg was coming their way. 'All of this is to say that towing an iceberg is not 100 per cent effective – depending on the conditions the berg will do its own thing.'

Following in his father's footsteps was not Bruneau's life plan. 'There's a lot of osmosis at the supper table', he told me, by way of explanation. It was also just a case of timing. In 1997, the younger Bruneau, who had been overseas for many years, found himself back in Newfoundland working on ice when James Cameron's *Titanic* was released, sparking a swell of public interest in the scene of the crime, Iceberg Alley. Suddenly reporters from the *Wall Street Journal* were turning up in Newfoundland asking about the bergs. Articles and documentaries about the threat of sea ice abounded; whale-watching guides expanded their season, taking tourists on day trips to marvel at these alabaster giants. Poems were even written, with one by Wallace Gagne about the berg that sank the *Titanic* ending: 'I wonder what has become of you. Is there any part of you still bobbing on the ocean; are you a fragment floating in a Martini in some expensive Ginza lounge . . .'

Icebergs were iconic, and Bruneau ended up being the go-to guy for information. He wrote a book about Iceberg Alley, securing his reputation as an iceberg expert. 'We are a decent city – I mean, small by international standards, but we have an international airport, a university, everything that makes a modern town. And yet we have these Arctic wanderers that get stuck in our narrows outside the harbour. People are fascinated by them.' Even with advances in technology, however, he doesn't think mega-bergs – some of which can weigh billions of tons – would survive a journey across the high seas. 'Problems cascade', he said. 'What I'm talking about is the staggering accumulation of risk factors. Every step of it has challenges', he said, relaying an imaginary conversation on the radio port: 'Yeah, failed to get the line today. We're off to a rough start, so all the ships are just at station and we're waiting for the weather to clear. That'll be tomorrow.'

'How many of those even happen before you've added a month to your travel time?' he asked me, rhetorically. 'You've got half of the ice you started with? And then you have a heatwave and the thing is vanishing before your eyes. By the time you get any ice or water into the water system, I don't think you're going to have 10 per cent of what you started with.'

Surely, the bigger question is whether we should even be considering this. 'If you're sitting there staring at a globe, it's easy to think that the problem isn't a lack of freshwater; it's that it's all in the wrong places. Look at all this white stuff at the bottom and top of the world. Why can't we just have some of that and we will all be just happy?' quipped Bruneau. But it's more complicated than that. 'Historically, there are so many things that we've tried to do for the best that have had unintended consequences', Geraint Tarling commented. 'You try to solve a problem in a way that you think will better humanity, but you just can't see the outcome.'

With A-68, the worst fear was the berg would become lodged off South Georgia, restricting ocean access for the island's wildlife. But this soon dissipated as the ice began to melt. Bit by bit, it calved smaller bergs off its mammoth structure, releasing 152 billion gallons of freshwater in the vicinity of the island. By April 2021, a trillion tons of ice had been released into the Southern Ocean. Scientists such as Tarling, at the British Antarctic Survey, are now trying to understand the impacts. The suggestion of towing icebergs across oceans to solve the world's water shortages may be a folly of sorts, but it's symptomatic of our relationship with the planet. As challenging as such a scheme sounds, to us it seems somehow easier than looking for durable solutions to our environmental problems, such as reducing water use in parts of the world, curtailing our greenhouse gas emissions, or consuming our ocean resources more mindfully. At a societal level, this inertia means we consistently choose the present over the future, taking ecological decline as the inevitable hit.

11

PARADISE LOST

For a coastal town on the Italian Riviera, Sanremo is – some would say – on the wrong side of Monte Carlo. Unlike its world-famous counterparts along the French Côte d'Azur, which boast Michelin-starred eateries and designer clothing outlets, the Italian town is known for its Japanese gardens, a twelfth-century cathedral and a nearby artists' colony. The main street is a good bet for snagging a simple souvenir or an affordable, if not an Armani-branded, espresso. Sanremo also happens to be a departure point for trips to the Pelagos Sanctuary, an offshore haven designed to protect some of the ocean's most majestic creatures, which is why I found myself there one July, not so long ago.

Summer in Sanremo is glorious; the weather is reliably warm, with a light breeze. But far out in the open ocean, strong northwesterly winds blow in, churning the deep sea and bringing to the surface plentiful food for the whales, dolphins and seals that make Pelagos one particular spot – in the vast ocean – worth protecting. Places like Pelagos are few and far between. And for the most part, they're completely open to the exploits of industry. That is especially true for those waters that lie beyond national governance. Pelagos was lucky, in some regards. More than a century ago, the reigning monarch of Monaco – Prince Albert I – recognised the significance of these waters and, in 2001, it became an official sanctuary, the first ever protected real estate on the high seas.

A few summers ago, I ventured to Sanremo to meet Nino Pierantonio, a biologist with a particular love of sperm whales. Nino works with Tethys Research Institute, a regional non-profit that fought for the creation of the sanctuary over twenty years ago and has since promoted the protection and conservation of its marine residents, largely through field-based research. Covering 33,750 square miles (an area roughly the size of Ireland), divided between the high seas and the coastal waters of Italy, France and Monaco, Pelagos became a sanctuary for marine wildlife: home to 8,500 animal species, and half of the Mediterranean's biodiversity, it is most famous for its sperm, fin and Cuvier's beaked whales, along with eight species of dolphin. Birds migrating south each year use these waters as a stopover, avoiding crossing long stretches of sea. The few days I spent surveying its waters left indelible impressions of long lazy sunsets enjoyed from our ship's deck, and the excitement of spying a misty spray of water in the distance, followed by an enormous, elegant shape rising up from the depths of a world we rarely visit.

MARINE SANCTUARIES AREN'T just pretty places, although Pelagos certainly is; rather, they are parts of the ocean that should meet a strict set of criteria, such as being rare, wild, vulnerable, lived in by extraordinary creatures and protected. Over the past twenty years, a number of formal, international processes have sought to pinpoint these areas, and have committed to putting in place a global network of protected areas that would stretch throughout the high seas, interconnected in a meaningful way, so that large mobile creatures like sharks and whales could travel between their breeding and feeding grounds while staying out of harm's way. One of these was an initiative of the UN's Convention on Biological Diversity, which in 2008 began a process to identify EBSAS (Ecologically and Biologically Significant Areas).

There are now over 270 EBSAS, including Pelagos, covering 14 regions. Some are in territorial waters, some are on the high seas, and others, such as Pelagos, straddle these boundaries. In a separate process, countries have identified parts of the high seas as 'Vulnerable Marine Ecosystems'(VMES). Many of these areas overlap with EBSAS but have been selected because of their fragility as well as their ecological importance. Under a separate scheme, we now have Important Bird and Biodiversity Areas – IBAS – and Important Marine Mammal Areas – IMMAS. Non-profits have made their own lists of areas considered high-seas 'treasures' and 'gems' or considered as being deserving of World Heritage designation. Mission Blue, a marine conservation organisation established by National Geographic Explorer Sylvia Earle – also known as 'Her Deepness' for her lifelong commitment to ocean conservation – has flagged seventy-six 'Hope Spots', special places that have been selected as being critical to ocean health. Put simply, there's no shortage of suggestions for where we should draw the boundaries of marine protected areas.

Implementing them, however, has been another story. Since Pelagos was created in 2001, eight other marine protected areas have been drawn up in high-seas waters: six of these form a network in the Northeast Atlantic – though intended to protect marine life there, industrial fishing is still permitted within these waters. The two remaining high-seas MPAS are both in the Southern Ocean, and are largely off-limits to industry. One, around the South Orkney Islands, is small, about the size of Pelagos, while the Ross Sea marine protected area covers 600,000 square miles, making it 1.5 times the size of the largest national park on land. A total of 432,000 square miles, or 72 per cent of the reserve, are strongly protected. It is considered our one credible effort to protect life on the high seas.

Pelagos, by comparison, represents a lot of what can go wrong with marine sanctuaries: intended as wildlife refuges, too often these ocean reserves end up being 'paper parks', places that are protected only in name. The criticisms most often made of marine sanctuaries are that they continue to allow for extractive industry, they have few regulations, they lack enforcement, and they are impermanent – their boundaries or status constantly up for negotiation. All of these are true of Pelagos. When the sanctuary was established in 2001, it was in response to lobbying by Tethys and other non-profits. Ostensibly an effort to protect its resident whales and dolphins, its legacy has been marred by its failure to keep these iconic inhabitants out of harm's way. Though Pelagos has never been blighted by whaling, even here humans have encroached on the welfare of these beloved giants: our presence has made these waters crowded, polluted, and overfished.

PIER B, NUMBER 22 was the instruction I'd been given by email, prior to my arrival in Sanremo. It seemed fairly straightforward, requiring little more than an hour-long train ride from Nice airport, and a short walk through the town's narrow, winding streets. Unaccustomed to the dizzying heat, which neared 104 °F, I took a couple of wrong turns, however, and found myself, an hour later than expected, lugging my suitcase down the concrete jut that was Pier B. Without clocking any of the numbers, I studied the line of luxury cruisers on offer – *Dangerous But Fun III*, *Shark in Love*, *Marilyn* (with a speedboat named *Tender to Marilyn*)

– before spotting the obvious misfit: a comparatively small, if elegant, wooden sailboat painted in blue and white with a cheery, lean Italian out front. Instantly affable, Nino is one of Tethys' chief scientists, and as such, he typically spends six to seven months a year at sea, much of it in the waters of Pelagos, where he has come to know the local whales on a first-name basis.

Born in Italy's mountainous north (nowhere near the coast), Nino has spent around eighteen years at sea, and at least ten in Pelagos. These days, he lives in the UK with his wife (also a conservationist) and their two sons, but he also spends long periods at sea collecting much-needed data on the whales and dolphins that inhabit Mediterranean waters. Nino wasn't always such an adept seafarer. As a student in Venice, he had planned to work on orangutans but, keen to get some field experience, he took an opportunity to volunteer with Tethys in 2003. 'Back then, I didn't even know that whales and dolphins lived in the Mediterranean', he told me. He soon became well versed in all of the cetaceans that inhabit these waters. But, for Nino, the sperm whale has had the deepest and most enduring impact. A single line drawing of a sperm whale covers his slender right calf. A Christmas present from his wife, it was sketched by a revered Berlin artist who positioned it perfectly so that when his calf flexes, the whale moves as gracefully as it would in the wild.

'People who are in love with the sperm whale call it the animal of superlatives. It's an animal that shouldn't exist', Nino told me. 'It looks like a dinosaur, with a head so large you could fit a van in there, or like a dragon in an animated movie, with its rippled bumpy skin and lopsided blowhole. It's a completely messy animal – it's extreme and fantastic in so many ways', he said, referring not only to the sperm whales' huge heads and big brains, but to the males' long periods of separation from their families – uncannily similar to Nino's own arrangement. Our boat, also named *Pelagos*, after the reserve, is in essence Nino's home away from home. For weeks every summer, he shares this space with a group of strangers, whom he welcomes on board with ease, exuberance and palpable excitement at having a fresh audience for his whale tales. Each volunteer pays around US$975 for the privilege, a fee that covers their accommodation on the ship, several hours of first-class tuition in cetacean biology by Nino, and a week's worth of delectable Italian food (including freshly brewed espresso and baked focaccia) made more impressive by the cramped kitchen, and more impressive still by the cramped kitchen in bad weather. These contributions are what keeps Tethys, and its research surveys, going.

I spent my week in Pelagos with Lorenzo, Mario and Lisa, three Italian marine biologists who served as Nino's research team, a young besotted, kissing Turkish couple, and three English schoolgirls intent on becoming either scientists or doctors (the trip was intended to help inform this decision). There were also two marine scientists in training – Anna and Tosca – and two return volunteers – thirty-two-year-old Marco, a convivial high-school teacher from Vercelli, Italy's rice-producing region, and Bernadette, a warm if shy Swiss HR consultant.

By all accounts, *Pelagos* – at 65 feet long – is a luxury Italian yacht, ideal for about eight people. There were, however, sixteen of us, which made for a fairly tight squeeze, especially with night-time temperatures over 85°F, no air conditioning, and a temperamental cistern. I spent the first night fretting about the growing global prevalence of extreme heat incompatible with human survival (and wondering if I was experiencing it firsthand), while each of my cabin mates relocated to the ship's deck to sleep outside beneath the stars, leaving me to nab the only bunk beside a porthole. Nonetheless, I quickly adjusted to life on board and over the days that followed *Pelagos* began to feel like home.

AROUND 5 TO 6 MILLION years ago, a cataclysmic episode shifted landmasses, squeezing the connection between the Atlantic Ocean and an adjoining sea named Tethys. This is how the Mediterranean Sea was born. A smaller and more sheltered version of the ancient Tethys Sea emerged from this seismic shift; at first it was little more than a deep, salty, empty basin, with little water flowing in, but it was soon breached, and flooded, by the Atlantic. It's now an oceanic island within the high seas. But for a narrow passage through the Straits of Gibraltar, it is cut off from the global ocean, which makes it rather special in terms of marine life. Though the Mediterranean covers only 1 per cent of the high seas in area, it is home to around 8 per cent of its plants and animals. Many of these are found nowhere else on the planet.

But the Mediterranean has another history too, one more familiar to us. The people who first settled along these coasts were strongly sea-orientated, taking full advantage of the rich bounty in the warm and salty waters that touched their shores. As far back as the eighth century BCE, the Phoenicians, renowned maritime traders, established trading posts along this entire coastline, from where they launched the earliest commercial sea fishery, capturing bluefin tuna that passed through the Straits of Gibraltar from the Atlantic Ocean back into the Mediterranean Sea. The Phoenicians also traded throughout the Atlantic and further still, with eastern cultures in Asia and also with Iberia in the west. They invented tools such as the first lead-filled sea anchor, examples of which still survive, and over time, developed the world's first maritime economy.

This seafaring outlook has endured, shaping the Mediterranean into a vibrant hub of life and commerce. Today, the Mediterranean Sea is bounded on all sides by crowded coastlines, criss-crossed daily by hundreds of vessels, exploited for oil and gas and fished heavily for food. An astonishing third of the world's marine traffic passes through the Mediterranean each year; cargo ships come from the Suez Canal and from the Straits of Gibraltar. In the summer months, passenger ships, day fishers and luxury cruisers all congregate along this coastline. A recent report, gauging the impact of noise pollution in these waters, listed 1,446 harbours and marinas, 228 oil drilling platforms, 830 seismic exploration activities, 7 million ship positions (from 2005 to 2015), and 52 wind farm projects.

In the midst of all this is Pelagos. Created with the best of intentions, it has its fair share of challenges.

These days, the sanctuary regularly makes it onto lists of global hotspots, but of the less desirable kind. Positioned close to some of the world's busiest ports, such as Genoa, Pelagos is a global ship-strike 'hotspot', a term that reflects the high number of whale collisions, often fatal, that happen here. Each year between ten and forty fin whales die in Pelagos from ship strikes. Sperm whales suffer this fate too, though in uncertain numbers. In recent years, 87 per cent of all whale-strike incidents within the Mediterranean have been within the sanctuary's waters. Pelagos has gained recognition as a plastics hotspot, rivalling the ocean's great garbage patches, and as a noise pollution hotspot, an inevitable outcome of being surrounded by all those harbours, ships, and oil and gas exploration exercises.

Worse still, perhaps, a 2016 analysis named the Pelagos Sanctuary as one of the three most heavily trawled marine parks in the world, finding that the sanctuary was fished by industrial trawlers for a total of 230,000 hours that year, a statistic that is surpassed in fishing hours only by a pair of marine parks along the Dutch north coast. Even though drift nets have been banned throughout Europe since 2002, they're still used illegally in Pelagos, and except for a ban on offshore powerboat races, the sanctuary has no rules to prevent any of this from harming its residents. According to a separate study, also published in 2016, and led by Italian researchers, the sanctuary is completely ineffective at protecting whales and dolphins in the Mediterranean. It's a view that's largely shared by conservationists.

And yet none of this is especially unusual for marine sanctuaries. Worldwide, there are more than 15,000 marine protected areas, and the vast majority allow commercial activity. Even in Australia's acclaimed Great Barrier Reef Marine Park, people can fish for endangered sharks, including hammerheads. The lenient designations are nonsense if you compare them to our history of protecting land. National governments started designating large reserves on land more than a century ago; Yellowstone National Park was established in 1872. Progress in the seas has been much slower. Ten of the world's largest MPAS, many as big as some countries, have been created in the past few years, spurred by political targets such as by a UN goal to protect 30 per cent of the ocean by 2030.

Large MPAS, even in remote regions, can make marine ecosystems more resilient. But what happens within the reserves is just as important as size. To be effective, MPAS must be strict. That means either no-take, where all commercial fishing is banned, or low-take, allowing only artisanal fishing for local consumption. Reserves this strict cover just 2.8 per cent of the Earth's seas, and 1 per cent of the high seas. To safeguard the seas, scientists say we need to sequester at least 30 per cent of the ocean in MPAS distributed worldwide, and that these must enclose a representative sample of marine life. Currently 75 per cent of marine protected areas globally is tied up in just forty-two large reserves, most located in remote tropical waters where little fishing or other industry occurs. Few MPAs target populated, temperate regions. Of more than 17,000 marine species studied in 2018, only about 500 had more than 10 per cent of their range within an MPA.

To have any discernible impact, marine protected areas must also be in place for at least ten years. Implemented properly, they can, however, bring enormous benefits for oceans and humans. If a region is truly left alone for long enough, fish and biodiversity can rebound even in places that were once decimated. More and larger fish spill over into neighbouring waters, too. Smarter MPAs can even make marine ecosystems more resilient to other pressures, such as pollution, warming and acidification. It's possibly too early to say how well the Ross Sea MPA is working, but there are some stellar examples from elsewhere.

One notable example is Cabo Pulmo, a coral reef close to the tip of the Baja California peninsula, and a place visited in the spring of 1940 by the American novelist John Steinbeck on a maritime adventure that would later inform Steinbeck's *The Log From the Sea of Cortez*. At the time of his visit, the reef at Cabo Pulmo, though small – about the size of Manhattan – was an underwater haven, the seafloor dotted with 'millions' of sea cucumbers, 'great numbers' of sea stars, 'knots' of brittle stars and plentiful sea urchins, as well as large sea snails of many varieties. So abundant was the marine life that settlers had first come to Cabo Pulmo in the eighteenth century and built up a town near the reef, from which they fished for mother-of-pearl shells.

Steinbeck, along with his friend and travel companion Ed Ricketts – an American marine biologist – documented the region's plant and animal life in *Sea of Cortez: A Leisurely Journal of Travel and Research*, which includes both their ecological observations and a catalogue of more than 550 species they encountered. Since then, Cabo Pulmo has undergone many changes. By the early 1980s, it had been so heavily fished that the reefs were destroyed by boat anchors and many of the larger fish were long gone. By 1985, the locals were so concerned that they agreed to a radical plan; they would stop fishing the reef. Somewhat amazingly, it worked. By 2009, Cabo Pulmo had recovered. Fourteen years after locals set up the 'no-take' MPA, scientists surveyed it. Much to their surprise, the fish had returned in great abundance – increasing in biomass by 463 per cent – as had large predators and underwater plants. Cabo Pulmo is now protected by UNESCO, the global body charged with preserving sites that are of cultural, historical, scientific or other significance, as one of a series of 244 protected islands and marine zones in the Gulf of California. Cabo Pulmo is testimony to the ocean's amazing capacity for restoration. Now, conservation scientists are hoping that the same can happen throughout the oceans, even in international waters that currently receive little protection.

But right now the goal of protecting 30 per cent of the ocean still seems a long shot. In recent years, under pressure to meet this target, national leaders have been scrambling to draw up MPA boundaries. In the rush, some have created weak reserves that have few restrictions. Other nations, including the UK, have created vast no-take sanctuaries around remote overseas territories, such as Pitcairn Island in the Pacific. These places, however, have no large commercial fisheries or other industry, so it is unclear if they offer much conservation benefit. In international waters, the political process of creating MPAS is in its infancy, so it's too soon to tell whether the same will apply, or whether we'll enforce stricter rules offshore. The challenge there will be not only agreeing to these rules, but monitoring MPAs that are out of sight and out of reach. The larger goal, of course, should not be maintaining the status quo, but restoring and rehabilitating marine ecosystems to their former glory, just like at Cabo Pulmo.

THE FIRST DAY of surveying Pelagos' turquoise waters with Nino and his group of volunteers didn't go as planned; after only a short time offshore, the wind picked up, rocking our sturdy little ship and turning the less seafaring among us 'mozzarella' white, according to our Italian hosts. We returned to port where we could stomach something other than dry crackers and gathered in the ship's mess to hear Nino deliver a movie-length talk about the whales and dolphins of Pelagos, a welcome feature of the days ahead. These lectures, often delivered at the end of the day or during a bad spell of weather, covered everything we needed to know about the cetaceans we'd come to see: their lifestyles, their habitat and their sounds.

I soon learned that Nino not only loves talking about whales, but also listening to them. Give him a hydrophone (an electrical microphone for underwater use) and he'll quickly tell you who is underwater and what they're doing. When he's at home in the UK, he likes to make his own hydrophones from bits of wires, metal and cardboard; by dropping them into a lake or on the shore, he shows his sons how to eavesdrop on another world. In Pelagos, he can differentiate each of the sanctuary's residents by sound alone. His favourite sound, unsurprisingly, is that of the sperm whale, an animal capable of producing noises as loud as a jackhammer.

On that first evening, we gathered in the ship's mess in sweltering heat listening to a variety of recordings that Nino had gathered over the years from various nearby locations. Among them was the low-bass sound of Pelagos' largest resident, the fin whale, which, to human ears, is virtually inaudible. This was followed by the long, high-pitched 'whistle' of a common dolphin and the short, peppy 'trumpet' of a male sperm whale, an occasional sound made just before they flip their tail and dive deep. Most familiar was the song of the humpback, a sound so strange that it is instantly recognisable, at least to anyone who has been to a wellness centre or on a yoga retreat. To Nino, it was 'terribly annoying, like a crazy, crazy cow shouting at the moon'.

Sperm whales essentially see the world through sound, and while they don't make the elaborate songs of humpbacks, they do have their own repertoire. 'In the cetacean world, sound is the tool to do everything; it's not just about seeing where they're going, they use it for orientation, for navigation, and to keep family bonds together; they also use sounds during migration, and for finding prey', explained Nino. Sperm whales are unique animals that, over millions of years, have accrued some bizarre adaptations to life in the deep. The largest of the toothed whales, they can grow to 60 feet long. They can dive to over a mile and stay there for over an hour. But the sperm whale's most impressive feature is its enormous head. Taking up a third of its body both in length and weight, it is the whale's sound generator, and the noises it produces – including clicks, squeaks, clangs, trumpets and creaks – are among the loudest in the animal kingdom.

The head of a sperm whale is an elaborate affair. At the front is the spermaceti organ, a large cavity that contains a series of oil-filled sacks, called the junk – a name that comes from the fact that after killing the whale and extracting the oil, these sacs, considered worthless, were thrown overboard. The oil of the spermaceti organ, however, was immensely valuable to nineteenth-century whalers, who used it in lamps and to make soap and candles. Reaching out from the spermaceti organ are two long nasal passages; the left connects to the blowhole and the right forms a little air-filled sac between the spermaceti organ and the frontal lobe. This sac can produce sound by forcing air into the right nasal passage and through to a massive valve called the monkey lips. When the monkey lips shut, the air is released and a sound is generated; this sound then bounces back through the spermaceti organ and is reverberated inside the whale's head, released into the ocean as minute bursts of energy.

'All this bouncing back and forth, back and forth, happens in very short time – around 200 milliseconds', said Nino. 'And all these little packets of released energy make up one single click. That's how we perceive it.' We huddled around to listen, just a bit harder, to the sounds emerging from Nino's laptop. As I leaned in, captivated by these ethereal sounds, I was struck by how disconnected we are, most of the time, from the subsea world; it's so rare that we have a chance to appreciate its beauty up close – another reason, perhaps, why we dismiss it all too easily in considering how we use, exploit and protect this space?

Most of the sounds of sperm whales are used for echolocation, and the most common type is for hunting prey, such as squid. As a sperm whale glides through the ocean, it clicks as a way of searching out prey in the dark – much the same as bats do in the dark – gathering information on food that might be as far as 10 miles away. Each click lasts just one to two seconds and is followed by another just half a second later. Once the sperm whale homes in on its prey, it switches tack, and starts to 'creak'. These sounds, much shorter in duration (lasting about 0.1 of a millisecond each), higher in pitch and more tightly packed, focus the whale's attention on its immediate surroundings. To us, this sounds like a creaking door. If the creak is followed by a long silent pause, the sperm whale has caught its prey. If it's followed immediately by another creak, it hasn't been so lucky.

And then there are the codas, which are like language. Used for communication, the sperm whale coda is a regional dialect or a sort of morse code that is understood by members of the same clan. For the most part, codas are used just by female and calf sperm whales, who stay resident in the same, warm waters year-round. Once they've bred, male sperm whales are migratory and often solitary, and so possibly don't have the same communication needs as the sociable females. In Pelagos, and throughout the Mediterranean Sea, sperm whales use a 'three plus one coda', which consists of three short clicks followed by a pause and then another: click click click – click. In some parts of the ocean, where there is more than one sperm whale clan, there can be different codas within the same region. In the Eastern Caribbean near the Virgin Islands, for example, there are two clans. One has a 'one plus one plus three' coda (click – click – click click click), while their neighbours use a '5R' coda (click click click click click), the R denoting regular intervals between each click. Even when two clans such as these overlap geographically, and they both use coda, they don't speak to each other. Having heard these sounds on Nino's recordings, I was eager to hear them for myself, in real time. If I listened closely enough, perhaps I could?

'SOFFIO, SOFFIO. Three o'clock, 500 metres.' Marco, the veteran whale watcher among our troupe, let out a series of short instructions telling us that he'd seen a whale's blow ('soffio' in Italian), located at a 3 o'clock position, roughly a quarter mile in front of us. Nine miles south of Monte Carlo, we had been covering the same stretch of ocean, back and forth, looking for a sperm whale that we spied briefly earlier. As the ship stalled, it rolled from side to side, and an oddly familiar whiff of diesel wafted onto the deck, filling my nostrils. Compared to the blue-black of the Arctic Ocean, the waters of Pelagos are ultramarine blue, their soft waves catching glimmers of the sun at every turn, making it hard to discern shapes small or large.

Suddenly, the whale surfaced, arched its broad back, raised its mighty tail and flaunted it in our direction, before gracefully retreating to the depths beneath. Barefoot and agile, Nino ascended the skinny ladder to the crow's nest to get a better view but returned to the bow just in time to photograph the fluke – the tail – which, for ID purposes, is as good as getting a mug shot. He grabbed hold of the ship's wooden railing and steadied his feet on the white fibreglass hull, beginning to talk excitedly, the sighting serving as a cue for another whale monologue – this time about how he identifies and catalogues the whales. We gathered at the ship's railing, near the bow, our feet dangling over the side, listening attentively while hopeful of spying a plume of noisy mist in the distance, followed by the colossal shape of a sperm whale. Half an hour later, it returned, performing its elaborate display for an enamoured audience all over again.

Below deck, Lorenzo, one of the marine biologists on Nino's team, deployed a hydrophone – a simple length of oil-filled tubing stretched out behind the ship – to listen for this whale and any others within earshot. Enthralled by the previous evening's entertainment, I was keen to join him. Nino suggested I go for a swim and stick my head, with its very own listening devices, around a foot and a half below. A scorching day, completely cloudless, we were all receptive to his idea. Marco led the charge in diving from the top deck, while I took a more cautious route down the ship's ladder. I dived in, forcing my head a foot and a half below the surface of the warm, briny sea. For a few brief moments, I heard a click, so faint I wondered if I'd imagined it. It was all too quick before I surfaced and so I grabbed hold of the handrail on our racing yacht, and plunged down again feet-first, staying there as long as my body would allow me before my brain took over, yelling at my lungs to grab some air. My eyes stung with salt, but I plunged again, until I was certain I'd heard something – perhaps half a hello?

These codas between whales, which I was lucky enough to eavesdrop on in Pelagos, form part of the soundscape that has filled these waters for millennia. But that soundscape is now changing.

In February 2021, an international group of scientists warned, in a paper published in *Science*, that in the past 200 years, we've made the ocean noisier by introducing shipping, industrial fishing, coastal construction, oil drilling, seismic surveys, warfare, mining and sonar-based navigation. In the past fifty years in particular, the noise generated along major shipping routes, such as through the Suez Canal and the Straits of Gibraltar, has driven marine life away from vital breeding and feeding grounds. Even noise from land – on bridges and at airports – can penetrate deep underwater. Accompanying the February 2021 paper was an audio recording, fashioned by multimedia artist Jana Winderen and intended to focus readers' ears on this shifting soundscape. The track begins in a quiet pre-industrial ocean. Soft, chirpy sounds, like birdsong in a country meadow, chime out, uninterrupted. Over time, these are overlayed with deeper, thunderous tones. In the background, an irksome cacophony of unidentifiable clangs and clatters gradually becomes louder until it dominates, drowning out the more jovial tones.

It's a similar background soundscape, a racket if you want to call it that, which is now annoying and stressing the ocean's inhabitants – not just the larger creatures such as whales, but everything from invertebrates upwards. Apparently, we are making so much noise that it's masking normal cues – codas between sperm whales, for instance, or opportunities to catch a prey or avoid a predator. On hearing this news, it struck me how quiet Pelagos must have been, 200 years ago before it became the hub of activity that it is now. Had I gone for a swim a couple of centuries previous, would I have heard a whole series of clicks, a creak or a full coda conversation?

Later that afternoon, on the ship's computer, Nino showed me a graph, called a spectrogram, that represented a recording he had made that day of one of the sperm whale's clicks. Zooming in to an individual click, we watched the discrete packets of energy, all of which add up to make one audible sound. By measuring the time taken for a single click, it's actually possible, Nino explained, to work out the size of the whale. From strandings, which offer a rare opportunity to autopsy and study a whale in detail, scientists now know the density of an average whale head and the average speed of sound through it. By recording the clicks, you can measure the time taken to make each one, and from there, you can work out the distance taken for the sound to travel, which is the width of the whale's head. Sperm whales typically have a head one-third the size of their body length, so by recording the clicks, it's possible to figure out the rest. 'With all the data we're collecting now, I'll be able to figure out how big this whale is. I'll tell you later tonight', said Nino, explaining that to get a reliable estimate, he'd need to analyse at least 800 clicks. 'How many clicks do you think you got today?' I asked. 'At least 3,000.'

THE FOLLOWING DAY, we ventured further offshore, to where Pelagos' waters become the high seas. Throughout most of the ocean, the high seas start at 200 nautical miles from shore, at the outer edge of a nation's Exclusive Economic Zone. The countries that border the Mediterranean Sea, however, had not claimed an EEZ at the time that Pelagos was created, meaning that international waters could be reached just 12 nautical miles from shore. From land, it's possible to reach the high seas of Pelagos in around two hours. The duration of the trip meant leaving the dock at 6:30 a.m. The sailing was smooth and so I sat at the ship's bow, my feet a short distance from the sun-kissed waters below. A couple of striped dolphins joined me for the journey, their heads bobbing up and down below the bow. After three hours of sailing, we were positioned in an area where whales typically come to feed, about 20 miles from shore, in waters about 165 feet deep. Having had our fair share of luck spotting sperm whales, we were hoping to see the sanctuary's physically bigger icon: a fin whale.

Within a half hour of our arrival, we spotted a series of blows, at 8 o'clock from the bow: a fin whale, the tell being its 'fluke print', a shimmering, still patch of water that it leaves behind as it dives. Outsized only by the blue whale, the fin whale is the second largest mammal on Earth, reaching lengths of up to 80 feet and weighing as much as 90 tons – about fifty times the size of an average car. Fin whales are easily recognised by two distinctive features: a notable ridge along their dorsal fin, which gives them the nickname 'razorback', and the asymmetry of colour in their lower jaw, with the right side being mostly white and the left side being mostly black. Unlike the sperm whale, which flaunts its fluke as it dives, the fin whale does something called a roundout, which gives you a better view of their side and the noteworthy dorsal fin. Over an hour, we tracked and watched three fin whales and one sperm whale, within 2 miles of our boat. The fin whales stuck around for longer, resting at the surface, ambling around then softly blowing, before gearing up for another dive.

As we headed back to shore, I retreated below deck to catch up with Nino, finding him working in the 'office', a small chair-sized cubby across from the galley. He was flicking through a catalogue of whale-fluke photos, each with a name attached, trying to match them with the ones we had just seen. For the sperm whales, he used the tail fluke to ID them, whereas with the fin whales, he used markings along their sides and the overall look of their dorsal fins. 'We use the general shape of the fluke, the trailing edge of the fluke and then all the marks and the notches. The missing bits. If there are scars, if there are marks, any of those things, and then we compare them', explained Nino. He plucked out a file with the name Alex marked on top. 'Wow, that's quite distinctive, isn't it?' he said, showing me a gorgeous pattern on its tail fluke.

We went through the photos one by one, comparing spots, notches and bruises. Nino called out the name on each photo as we went, trying to ID one particular fin whale we had just seen. 'Charles, no, Chipper, no. I actually called this one – when I saw him the first time – Mirror. They changed it', said Nino. 'This one is called "Clean" because there are no markings. Comb? Dot? Dave? Devil? Diego? Doug? Drascoe? Eliot? All of the whales ever photographed in Pelagos are catalogued alphabetically; it's part of a protocol they follow at sea that follows internationally accepted standards. There was one called Nino, but Nino's favourite was Luke. By cataloguing these animals, they can accrue very useful information, about their movements and preferred locations. 'Eric? Yes. It's Eric. There you go. Look at the general shape here. Eric. There he is', said Nino excitedly. 'They don't have to be beautiful pictures. They have to be functional', he said, showing me an image of a fin with a chunk missing – reminding me with a painful thud why I was there in the first place. Not all, but many of these features, the scars and aberrations, were the effects of a ship strike.

IN NOVEMBER 2020, the Italian architect Angelo Renna launched his latest research project, Sweep Island. Known for his conceptual projects, Renna previously designed a 300-foot-high artificial mound for the Italian city of Turin known as 'Sponge Mountain' which would serve both as a city park and a natural means of carbon capture; another is 'Rub the Belly', a multipurpose laboratory and climbable landscape made from recycled rubber. Sweep Island is intended to help mitigate Pelagos' plastic pollution problem while also giving its marine life a helping hand. An artificial habitat, Sweep Island has a top made of soil, trees and bushes – the sort of place that Pelagos' overwintering birds might care to pay a visit to, or that might harbour a variety of insects. Below the water, the island offers another unique setting: made of troughs and peaks, and with lots of hidey-holes, this diorama is designed to amass algae and plants, attracting invertebrates and fish, much like a coral reef. Stretching deeper still – down to 15 feet below the surface, the depth at which most plastic waste gathers – is 'the collector', an intricate 3D shape designed to intercept plastics and keep them out of harm's way.

For the moment, Sweep Island is just a concept, and it's impossible to know just how many of these you'd need to clean Pelagos of its plastic. Hundreds probably, but Sweep Island has nonetheless drawn attention to one of Pelagos', and the ocean's, most intractable problems. Each year, between 4.4 and 13.2 million tons of plastic waste enters the ocean. Most of this, around 80 per cent, comes from land, while the remainder is disposed of at sea; some in the form of abandoned and lost 'ghost nets'. These large plastics – cups, milk cartons, bags – are gradually eroded by wind, waves and sunlight until they become no more than tiny specks drifting on the currents. Along with the mini beads of plastics found in many cosmetic products and the microfibres that shed from synthetic clothes, these 'microplastics' are now found throughout the world's ocean, littering coastlines and the open ocean.

With over 3 trillion microplastic particles estimated to be in its waters, the Mediterranean is the most polluted sea in the world, and Pelagos is its worst affected area: reaching 3 million pieces per square mile, plastic now gathers here in concentrations that rival those found in the ocean's garbage patches. The smallest fragments – which can be anything from 5 millimetres down to 10 nanometres in size – pose a particular threat to the sanctuary's visiting seabirds, while larger pieces – including plastic bags, straws, and so on – are ingested or cause entanglement for larger animals.

In Pelagos, micro-debris and plastic additives accumulate in even the smallest of food items – plankton – which in turn become contaminated food for fin whales. It's the larger, and more visible, plastics that cause most public concern: in the past few years, images of dead whales, stranded and with bellies full of plastic, have made headlines the world over. In May 2019, the carcass of a young sperm whale, seven years old, was retrieved from a beach in Cefalù, Italy, its stomach filled with plastic. The previous month, a pregnant sperm whale became stranded in Sardinia with nearly 50 pounds' worth of plastic bags, containers and tubing in her stomach. While Sweep Island won't solve this problem, there are other practical efforts underway. In 2015, for instance, the Prince Albert II of Monaco Foundation joined forces with other environmental groups including the non-profit International Union for Conservation of Nature (IUCN) to launch the Beyond Plastic Med (BeMed) initiative, which to date has funded fifty-three projects in fifteen countries, with the aim of reducing plastic pollution at source, improving waste collection systems, raising awareness, collecting data and implementing new regulations. An advisory board of scientists is at hand to ensure the effectiveness of proposed solutions.

It was four days into our whale-watching adventure before we came across our first sizeable pieces of plastic. The first was a flat sheet of off-white plastic, which I had mistaken as a manta ray. Later that same afternoon, while I was admiring the childlike energy of a pod of Risso's dolphins porpoising near the ship's bow, Marco spotted an inflatable crocodile drifting past. Paulo, our skipper, quickly turned the ship around and positioned it so that we could hook it from the bow. We hauled it aboard, named it Elton and lashed it to the stern in a spot that we had allocated to plastic debris. The following day, Elton was joined by a couple of jerry cans and another inflatable, this one more worn and designed to look like a tray of pink cupcakes.

I asked Nino about the effect all this plastic waste is having on Pelagos' whales and dolphins and whether scientists know, from necropsies, whether it's actually killing them, or if they are dying of other causes. On the positive side, he answered, there's been a noticeable decrease in the number of animals dying in plastic drift nets. 'This EU ban is working', he said, referring to the fact that drift netting had been banned since 2002. There are still isolated cases, he said. 'But they are very, very few now.' With ingested plastic though, the picture gets more complicated.

Fin whales like to spend a lot more time on the sea surface than sperm whales. They do something called logging, which is essentially just resting at the surface. Before any single fifteen-minute dive, a fin whale typically logs for fifteen minutes.

A sperm whale, on the other hand, usually spends just a third of its time at the surface, which means it's less likely to be the victim of a ship strike. Ingesting plastic, however, can change that. Sperm whales that have taken in a lot of plastic typically alter their behaviour, spending more time at the surface, presumably in distress or discomfort. There, they're more susceptible to ship strikes and if they are hit and stranded, the post-mortem will declare the animal killed by a vessel. 'So, the primary cause of mortality is the ship strike, not the plastic industry', said Nino. 'We don't know at the moment if there is a direct connection between ingesting lots of plastic and the increasing number of ships striking sperm whales, but it's an interesting question.'

ANYONE CAN DRAW a line around a part of the ocean and call it a sanctuary. What matters is what happens within those boundaries. Conservationists and scientists agree that to protect marine life, we need to place 30 per cent of the ocean within sanctuaries by 2030, and use the rest sustainably. They also agree that for marine sanctuaries to be effective, they need to be no- or low-take, meaning that they exclude industrial fishing; ideally they should also outlaw other extractive industries such as oil and gas exploration or seabed mining. On some level, Pelagos' status – as a failed marine sanctuary – beggars belief. Why establish a sanctuary only to maintain the status quo? But the fact is that most marine protected areas do little to safeguard marine life, and so, in that context, Pelagos is not outside of the norm.

On an international level, there are a plethora of processes for creating MPAs. Drill down to the regional or local level, and you find a whole other layer of protective legislation. In the Mediterranean, a complex tangle of interwoven and overlapping legislative agreements, bodies and sites exist all with the aim of managing or protecting the region's biodiversity. For a start, there's the General Fisheries Commission for the Mediterranean (GFCM) and the International Commission for the Conservation of Atlantic Tunas (ICCAT), which are supposed to manage fisheries. Overseeing conservation is the Agreement on the Conservation of Cetaceans in the Black Sea, Mediterranean Sea and contiguous Atlantic area (ACCOBAMS), and the Barcelona Convention. This latter agreement has identified thirty-five Specially Protected Areas of Mediterranean Importance (SPAMIS) within the region, proposed by ten different countries. Of these, only Pelagos extends to the high seas. Yet in spite of these efforts, the prevailing view is that we're failing to protect marine life in these waters.

But what does that mean for Pelagos, a largely lawless place that has been designated as a marine protected area? Pelagos is politically tricky, in part because it can be difficult to rewrite an established agreement, especially one that's been forged between different jurisdictions – in this case France, Italy and Monaco. One option is to discount it entirely and start again elsewhere, hoping to protect these species in other parts of their range. Perhaps further offshore where there are fewer competing interests and fewer jurisdictions to manage? Another view is that rather than designating Pelagos as a strict reserve, managers could implement some specific rules to tackle the worst offenders.

As of June 2021, the World Wildlife Fund began formal preparations to apply to the International Maritime Organization (IMO) – the UN body that oversees shipping – to designate Pelagos and surrounding waters as a Particularly Sensitive Sea Area (PSSA); if it qualifies, then the IMO might agree to adopt rules that would slow ships, reroute them to avoid important feeding and breeding grounds for whales, and implement better vessel reporting and maritime traffic management. Given the difficulty in creating no-take sanctuaries, this pragmatic approach might offer hope for other places, where outlawing humans seems impossible.

As we headed back to port on our final evening surveying Pelagos, we reflected on a good day's work. From the top deck, scanning the horizon with a pair of binoculars, Lisa, one of the scientists in training, had bagged the season's fifty-ninth sighting of striped dolphins and a ray. Earlier that day, she had also spotted an ocean sunfish (*Mola mola*) sunning itself lazily not far from the ship's stern. The heaviest of all of the bony fish, the *Mola mola* is a huge, vertically flattened creature that can be longer than an adult human male and can weigh in at over 4,400 pounds.

This *Mola mola*, just a few feet from the ship's bow, was engaged in typical sunfish behaviour: lying on one side and catching the sun's rays, most likely after a deep dive, in search of food, in the ocean's icier depths. As the others danced around Elton the inflatable crocodile while 'Loser' by Beck blared on the radio, I went in search of Nino and found him fiddling with a small blue toy octopus in the wheelhouse, chatting with Paulo the skipper, an ex-fisherman. I asked him what he thought the chances were for turning Pelagos around, and whether it was even possible. 'I've always been a big fan of the Pelagos Sanctuary', he said. 'But', he added: 'Pelagos was set up more than twenty years ago, and since then, the thinking has evolved on how we protect species, how we protect environments, on what actually is conservation and protection.' With all our new-found wisdom, shouldn't we make Pelagos a strict reserve, with industry outlawed? Nino wasn't sure that was either possible or desirable. His view is that we can't make rules that treat marine life as though it exists separately from its surroundings. 'We have to preserve the environment and species, but also the cultures and traditions that exist here', he said. 'We can't pretend this place exists in isolation.'

Maybe it is too late to turn Pelagos around, and transform it into a true sanctuary. But I wondered – given the pace of our offshore encroachment, and our failure, thus far, to protect even this small portion of the high seas – was there even any hope that we could achieve this elsewhere? It's not possible to change the past, to alter hundreds of years of history, but perhaps we can chart a different course for the future.

12

HOPE FOR THE HIGH SEAS

IT'S EASY TO lose hope for the high seas. But there are reasons to remain optimistic. Chief among them is a forthcoming treaty, agreed by UN member states in March 2023, to protect the half of our planet that is ocean beyond national boundaries. The idea of a new rulebook for the high seas was first conceived of in 2001, when a group of conservationists gathered on a small island, named Vilm, in the Baltic Sea. Located just off the north German coast, this tiny islet first emerged from the waves some 6,000 years ago as a heap of debris left behind by a departing glacier. In the intervening millennia, visitors have come and gone. Artefacts retrieved from the island suggest it was first discovered by Stone Age people, and later by Slavics, who built a temple. More recently, in the nineteenth and twentieth centuries, the island became a summer home for aristocrats and for high-level politicians. Lacking permanent habitants, Vilm has remained remarkably unspoilt. There are no televisions, shops or commerce of any kind, and this miniscule mound – smaller than half a square mile – boasts some of the most pristine forests in Europe, containing over 500 plant species, as well as an abundance of birds and small mammals. Its few cottages now welcome conservationists, scientists and artists who wish to study or appreciate nature. Vilm serves as a gentle reminder of how bountiful the natural world can be in humanity's absence.

It was here, in February 2001, that a group of thirty-three biologists, conservationists and lawyers first met to discuss what to do about the ongoing damage to the high seas. One of those experts was Kristina Gjerde, a high-seas specialist with the International Union for Conservation of Nature (IUCN), the global body charged with assessing and protecting the natural world. She recalls the meeting vividly, not least for the freezing weather (at that time of year, it's possible to walk on ice from Vilm all the way to the neighbouring island of Rügen, although most people choose the local icebreaker). What Gjerde remembers most is that in Vilm, she, and the other experts, saw for the first time video footage of the startling devastation being wreaked by bottom trawling, the type of industrial fishing that I myself witnessed later that year first-hand in the Northeast Atlantic. Gjerde and her peers watched in horror as fishing vessels dragged heavy metal chains and several-ton trawl doors across fragile coral reefs and over seamounts off Australia, in the icy fjords off Norway and in the depths of the Northeast Atlantic Ocean. In a matter of minutes, these trawlers had wiped out coral reefs that had taken thousands of years to accrete, leaving crushed moon-like landscapes in their wake.

At that point, in 2001, trawling was responsible for 80 per cent of the annual global catch of bottom-dwelling species, including prawns and orange roughy. Most of that was happening out of sight in the deep waters of the North Atlantic Ocean, the southwestern Indian Ocean and the Southwest Pacific Ocean, and with no law specifically protecting life on the high seas, it was perfectly legal. That year, the EU took 60 per cent of the global high-seas bottom-trawl catch and Spain alone took 40 per cent. The catch was incredibly small, less than a quarter of a per cent of wild fish landed, by weight, in 2001, and only enough to support 100–200 vessels year-round. But it was having an oversized impact – scientists had, by then, begun to document 95–98 per cent losses of coral cover from trawled seamounts. They had begun to witness the serial depletion of deep-water fish stocks.

It was then that the thirty-three experts assembled in Vilm began to envisage a new law to protect marine life from bottom trawling – and anything similarly destructive – in international waters. They formed an advocacy group, the Deep Sea Conservation Coalition, aimed at prohibiting fishing vessels from trawling fragile ecosystems on the high seas. Five years later, in 2006, they won that fight and a new rule came to pass, demanding that nations ensured their bottom-trawling fleets steered clear of vulnerable habitats, including corals and seamounts, on the high seas. In 2011, the Deep Sea Conservation Coalition launched a wider advocacy group, the High Seas Alliance, in preparation for a larger battle. Composed of the same core people, including Kristina Gjerde, New York lawyer Peggy Kalas, oceans advocate and consultant Matthew Gianni, and environmental lawyer Duncan Currie, among others, the alliance brought together twenty-three non-profits plus the IUCN to focus on pushing for a treaty that would protect marine life on the high seas broadly, by creating sanctuaries and putting in place stricter rules on how industry can operate offshore. In 2012, the year after its formation, the High Seas Alliance urged world leaders at the Rio+20 summit to back their proposal for a high-seas treaty. The plan worked and the UN was urged to move forward.

ON 4 MARCH 2023, their vision became a reality, when the High Seas Treaty was finally passed by UN member states in New York. It had taken twenty years of sustained effort, and six years of formal negotiations in New York (several of which I attended), to get there. The treaty has now been signed by more than eighty nations, but still needs to be ratified – and passed as domestic legislation – by at least sixty of these nations for it to become law. It might be years before its policies are implemented, but nonetheless the deal is a major win for conservation, the first international law to offer meaningful protection to the half of our planet that is the high seas.

For the deal to come off, nations had to concur on a few crucial issues. The first major focus was agreeing a framework to enable a network of marine protected areas (MPAS) to be created in international waters. Ideally, these would be interconnected and representative of the diversity of marine life and habitats on the high seas. The treaty would give UN member states the authority to propose new MPAS on the high seas, and a science-led council would then evaluate, and approve or reject, the suggested sanctuaries. This part of the deal was seen as crucial in giving nations a fighting chance of coming good on international promises to protect 30 per cent of the ocean by 2030.

Second, UN member states needed to agree a process for assessing the potential impact of new activities in international waters. There is precedent here. Currently, on land and in coastal waters, new commercial activities have to undergo an 'environmental impact assessment' or EIA, to weigh up any benefits against potential harm to wildlife and to ecosystems. On the high seas, only some activities are regulated in this way. It wasn't until 2006 that bottom trawling needed an EIA, for instance, thanks to the work of the Deep Sea Conservation Coalition. Even now, activities such as mesopelagic fishing, open-water farming and splashdowns of space debris do not need to consider their potential environmental harm. Conservationists want to see emerging activities – climate intervention or even sea-steading – on the high seas tightly regulated. A new treaty would make it mandatory for these activities to undergo a strict process of environmental assessment before getting the green light.

By far the most contentious issue that nations had to grapple with was how to regulate the use of marine genetic resources, meaning ocean creatures themselves and their gene sequences. This part of the deal was fraught with tension, partly because marine genetic resources in international waters are, arguably, owned by no one – or everyone. The deal that was agreed in March 2023 was a victory for conservation and for equality, for several reasons. First, it put in place a legal framework for creating large marine protected areas offshore. And while the aim is that these MPAS are agreed by consensus, if necessary they can be agreed by a majority vote, a strategy that averts the possibility of one major fishing nation blocking their implementation, as so often happens in Antarctica under the Commission for the Conservation of Antarctic Marine Living Resources' governance. Second, the deal mandates that new activities offshore are subject to an EIA; while that does little to change the status quo of resource extraction on the high seas, it makes it possible to slow its future diversification and progress. Lastly, the deal closes the growing inequality gap between the global north and south, by recognising that high-seas genetic resources belong to – and should therefore benefit – all of humanity. To this end, the treaty mandates for benefit-sharing, both monetary and otherwise, from the use of genes extracted from high-seas organisms. In theory, a portion of the profits from their commercialisation, potentially including the development of products worth billions in sales, will now be channelled back into a fund to protect the high seas.

While sharing the spoils of genetic resources was the most contentious issue for states to agree on, it is not where this new law will flex its muscles. Its power lies in its ability to protect, as much as is possible, the wonders of the high seas – those places far offshore, such as the White Shark Café, Lost City, the Costa Rica Thermal Dome, the Gakkel Ridge and countless others. Currently just 1 per cent of international waters are protected. That number is now set to grow, ultimately stemming the loss of biodiversity and helping maintain the health of our ocean.

But the High Seas Treaty only goes so far. One of its provisions, and a crucial compromise in getting it passed, is that it can't undermine existing authorities with jurisdiction over the high seas. That includes the International Maritime Organization (IMO), responsible for shipping, the International Seabed Authority (ISA), which oversees deep-sea mining, and seventeen regional fisheries management organisations (RFMOS) tasked with managing fisheries in various parts of the ocean. In practical terms, that means that the treaty cannot scrutinise activities already underway by insisting, for example, that they adhere to stricter environmental protocols. Included here is deep-sea mining, which is in an exploratory phase and is managed by the ISA. In reality, that means that so many of the issues raised in this book – overfishing, pirate fishing, climate change, dumping and pollution, deep-sea mining, Arctic shipping – won't be addressed by the High Seas Treaty.

There are, however, other efforts underway to tackle some of these issues. On an international scale, 200 UN member states have committed to agreeing a legally binding treaty to tackle the global plastic epidemic by 2024. The deal, which will address the full life cycle of plastic, including its production, design and disposal, should make serious headway in curtailing the production of single-use plastics, and the amount of plastic waste ending up at sea. On a practical level, researchers worldwide are focusing on technologies, including plastic interceptors that pull plastic waste into containers and out of the sea, as well as plastic-munching enzymes and bacteria, to clean up plastic waste from places such as the Great Pacific Garbage Patch. Meanwhile, the Port State Measures Agreement – which came into effect in 2016 – is the most significant major international effort to reduce pirate fishing; as more nations sign on, there will be fewer ports in which fishers acting illegally and under the radar can land their catch.

In 2022, the World Trade Organization moved one step closer to demanding greater accountability from governments in how they subsidise their fishing fleets. WTO, however, ultimately failed to reach a deal on ending the sort of harmful subsidies that facilitate unsustainable and illegal fishing on the high seas, but the hope is that the process now underway will continue to strengthen over time. Elsewhere, the UN Framework Convention on Climate Change has continued, since the Paris Climate Accord was signed in 2015, to urge nations to make deeper commitments to limit global warming to within 1.5°C (2.7°F) to 2°C (3.6°F) above pre-industrial temperatures. Right now, nations are failing to meet their commitments, with dire consequences on land and in the ocean; unless we change tack, we'll almost certainly need to rely on large-scale technological intervention to reduce climate impacts on humans.

Of all these efforts, the High Seas Treaty is humanity's first serious attempt to challenge the carnage that pervades the high seas. We must now hope that the foresight that empowered nations to agree this deal will translate to their broader behaviour offshore. In redressing our 'out of sight, out of mind' relationship with the oceans, the High Seas Treaty might just urge us to rethink how we use our ocean commons in ways that benefit the majority.

FROM THE BEGINNING, I always imagined I'd finish this book with an uplifting vision of the future, describing how the high seas will look, decades from now, managed properly. In this vision, there are sprawling marine sanctuaries filled with an abundance of animals. They are placed strategically, covering hundreds of thousands of square miles of our most precious places offshore – Lost City, Sargasso Sea, Gakkel Ridge, the White Shark Café, to name a few. Elsewhere, sustainable, ethical trades have emerged: seaweed harvesting, managed by sea-steading communities who power their high-seas homes with wind and wave energy, is one of them. Gene harvesting from the high seas is proving a boon to both the global north and south. A new anti-cancer drug, for instance, has recently been developed from the scaly-foot snail, which narrowly escaped extinction when deep-sea mining was outlawed internationally, deemed too destructive to proceed. You get the idea. Above all, this narrative was intended to reassure myself and others that, despite our fraught relationship with the ocean, everything will work out just fine. Won't it? Doesn't it always? But as I sat down to write, I realised that my view of the future isn't quite so rosy. And in spite of nations agreeing to a High Seas Treaty – which offers a real glimmer of hope for a different future – that, unfortunately, remains the case.

In part, I embarked on this project to counter my growing pessimism and anxiety at how our relationship with the natural world is evolving. Over the years I covered climate change intensely, first as an online editor at *Nature* and later as the chief and founding editor of *Nature Climate Change*; I watched as the levels of CO_2 in our atmosphere increased, year after year. It happened in spite of us fully understanding the causes of global heating and in spite of an abundance of political pledges to slow, if not solve, the problem. I reassured myself that there were other reasons to be hopeful; that if we couldn't come good on pledges to rein in greenhouse gas emissions, we could prepare for the changes ahead by making better choices elsewhere – by protecting large swaths of our planet and its inhabitants, by tackling other forms of pollution such as plastics, by refusing to allow new destructive industries to emerge. I realised the potential for hope in the high seas, that unclaimed part of our planet, which – if we're talking about volume of living space – is 95 per cent of Earth. As our largest global commons, with untapped resources, its fate still hung in the balance. I understood that while the high seas is no longer a pristine, unspoilt environment, it is Earth's last wilderness, and as such, it represents a sizeable opportunity to embrace our role as responsible stewards of this planet.

In writing this book, my aim was to impress upon people the crucial juncture at which we find ourselves. I wanted to strike a balance between conveying the damage we've already inflicted on the ocean, even far offshore, and the insights being gained from new and exciting science that might help us to forge a more sustainable way ahead. I deliberately sought people and places to visit that would help me tell those stories. In meeting Bjorn Bergman from SkyTruth, for example, I came to understand the scale of pirate fishing on the high seas, but also the incredible efforts underway to visualise, and curtail, the problem. In spending time at sea with Henko de Stigter, I better understood the risks from seabed mining and how science could inform us as to whether this nascent industry could ever be sustainable. In travelling to the far ends of the high seas, I grasped the pace of physical change underway, unprecedented in human history, in these remote parts of our planet, and the fresh challenges they present for conservation. I also came to realise how labels can be meaningless, and how 'protected' areas can be among the most exploited on Earth.

At the start of this journey, I, like many, viewed the high seas as a lawless frontier. I thought, somewhat naively, that if we just implemented more rules, we'd get a different outcome. What I discovered is far more troubling. For the high seas is far from lawless; as a space, it is highly regulated, if by a mish-mash of organisations and bodies, each using their own rulebook. I now realise that many of those tasked with governing this space wilfully ignore science and disregard expert advice. Time and again, they prioritise economic gain over sustainability – by progressing seabed mining in full knowledge that it will cause irreparable environmental damage, by increasing quotas for fish stocks that are in decline; by allowing extractive industries to operate inside the boundaries of marine reserves. Every time this happens, it invalidates the rules. The ISA will, no doubt, eventually have a mining code, with some requirements for prospective miners to carry out an environmental impact assessment. But it will be led by commerce as much as by science, and it won't stall the arrival of this destructive industry into the deep sea. The second problem, I've realised, is that on the high seas rules are, seemingly, made to be broken. Entire industries find legal loopholes – such as flags of convenience – so that they can do as they wish without penalty. Pirate fishers use slave crews to catch endangered species, tankers sink and spill their oil, and it's unclear, on the high seas, who exactly is to blame. Was it the captain from China, the owner from Spain, the London insurance broker, or the flag state, Panama? We have established the structures that allow this to happen.

But what does all of this mean for the future of the high seas? We're at a point in history where knowledge and technology are allowing people to conceive new agendas for the unclaimed half of our planet, ranging from intervening with climate change, finding new drugs and mining precious minerals. We can continue going ever deeper and further offshore in our quest for new sources of wealth, or we can strike a more sustainable balance. If we choose the latter, we need new rules; by agreeing to the High Seas Treaty, nations have made a real start. But for these rules to be powerful, they must be enforced and consistent. Right now, our ambition to develop economically continues to impede conservation, with the outcome that our laws are often hopelessly conflicted: the EU in 2021, for instance, continued to offer fuel tax relief to its fishing fleet while simultaneously negotiating with the WTO to end harmful fisheries subsidies that encourage overexploitation of stocks. Furthermore, the EU included a provision in its European Maritime, Fisheries and Aquaculture Fund (EMFAF) for 2021 to 2027 that provides €6.1 billion (US$6.6 billion) in subsidies for fishers to upgrade their vessels and travel further offshore. To anyone watching closely, these policies run counter to the EU's promise to abolish harmful subsidies and commit to sustainable fishing. Similarly, nations regularly commit to climate goals while pushing developments that increase their carbon emissions; Saudi Arabia is a case in point. Having committed to building a 'blue economy' from ocean-based carbon capture – with a goal of being net-zero by 2060 – the superpower's energy minister Prince Abdulaziz bin Salman recently announced Saudi Arabia's intention to be 'the last man standing' when it comes to oil extraction. Likewise, some of the nations most likely to ratify the High Seas Treaty, and commit to its vision of ocean protection, are among those nations that will continue overfishing and polluting the high seas.

Over the past few years, there's one question I've wrestled with almost constantly: what influence do any of us, as individuals, have to steer the ship on a different course? Much of the time, I'll admit, I feel as powerless as a piece of flotsam drifting on the currents, capable only of responding to the immediate choices in front of me. In my personal life, these are often simple choices, made in a fleeting moment between a busy schedule of work and parenthood – choices about where my food comes from, for instance, or whether to upgrade an ageing laptop. Our impact on the ocean is now woven into the fabric of daily lives to such an extent that we as individuals cannot stop the 'Blue Acceleration'.

That's not to say we can't play a role; we can, of course, make sustainable food choices, forgo eating fish caught on the high seas, we can reduce our carbon footprint, or use less plastic. But to make impactful change, we need new policies. We need to create new laws that don't simply promote extraction by dividing ocean space into exploitable units and resources, divorced from the ecosystems in which they are found. We need leaders who understand, and support, the need for evidence-based decision making. In addition to creating marine protected areas that could, eventually, cover 30 per cent of the high seas, we need to consider what happens in the remaining 70 per cent of ocean space. We need industries, including mining, pharmaceuticals and fisheries, that strive for accountability, ethics and transparency in their supply chains. As citizens and voters, we should support policymakers who are serious about, and take action towards, protecting 30 per cent of the ocean by 2030, in meaningful ways, and who do everything in their power to ensure the successful implementation of the High Seas Treaty that has now been agreed. We should support the non-profits that push for better transparency and accountability, such as Global Fishing Watch and the Marine Conservation Institute, as well as those that fight for more effective protection of ocean life, such as Greenpeace and other members of the High Seas Alliance. We can protect the ocean further by supporting effective legislative action on climate change. That may involve supporting calls for a carbon tax in your home country, and support for world leaders who fully endorse the Paris Accord. In reporting this book, I was surprised to discover that some of the most destructive high-seas industries, both existing and emerging, benefit very few. To the extent that we can, it's incumbent upon us, as individuals, to freeze out those practices and businesses.

More than ever, I feel that we need to be braver, to reimagine our relationship with our global commons. One possibility is closing the high seas to fishing, an idea that, in recent years, has been championed by scholars and conservationists. In 2015, Rashid Sumaila, a fisheries economist from the University of British Columbia, and National Geographic Explorer Enric Sala, among others, subjected this idea to a rigorous analysis. Firstly, they showed that less than 0.01 per cent of commercial fish landings, by quantity and by value, come exclusively from the high seas, so putting an end to those catches would have little economic impact. Many species targeted on the high seas, such as tuna, swordfish, cod and mackerel, are 'transboundary stocks' which move between EEZs and the high seas. Closing the high seas would, they maintain, increase these transboundary stocks to the extent that catches within national EEZs would increase by at least 18 per cent. The result of this is that the fish would be more evenly shared; rather than being monopolised by large distant-water fleets from China, Japan and Spain for instance, fishers from other nations, such as Angola, Greenland, India, Ireland and Britain would have a better chance at catching these fish. This wouldn't change the overall global catch, but it would cut in half income inequality from fisheries among the world's maritime nations. A separate analysis, published in 2014, claimed that closing the high seas to all fishing would increase the profit from fisheries by more than 100 per cent, increase the yield by more than 30 per cent, and improve fish stock conservation by more than 150 per cent. In short, closing the high seas could boost fisheries, giving average fishers better livelihoods, while helping us achieve our conservation goals.

If this seems radical, others have gone further still. The famed American author and naturalist E.O. Wilson spent the latter years of his life campaigning for 'Half-Earth', the idea being that to stabilise biodiversity loss, and ensure our own long-term survival on Earth, we need to set aside half the planet, or more, in a reserve that is off-limits to industry. In honour of his idea, we now celebrate Half-Earth Day in October each year. Wilson intended this 'half' of Earth to be spread out among the most biodiverse places, but a variation on this theme could be closing the high seas to all extractive industries. As taken as I am with these bold ideas, I also accept that, in the current environment, they are politically untenable. Nevertheless, they do hold value in opening up the conversation about how we manage and share the space that is the unclaimed ocean.

Positive change is possible, but it's hard won. It took twenty years of relentless work by conservationists to push through a High Seas Treaty. Other rays of hope come from the recovery of marine species through conservation efforts. A ban of drift nets longer than 1.5 miles, which came into force in 1993, is noteworthy. By outlawing the use of these fishing nets known as 'walls of death', we've managed to reduce the catch rates of non-target species such as sharks, dolphins and turtles. In recent years, there has also been a return, in response to a global ban on their exploitation, of humpback and fin whales, in historic numbers, to the Southern Ocean. As a result of stricter quotas, at least five of the ocean's tuna stocks are rallying.

In coastal waters are other examples where conservation efforts have borne fruit. A 2008 ban on tributyltin (TBT), a chemical used to keep the hulls of boats free from encrusting sea creatures, reversed an alarming trend in gastropod sea snails becoming hermaphroditic. Similarly, tighter safety regulations led to a fourteen-fold reduction in large oil spills from tankers from 24.7 events per year in the 1970s to 1.7 per year in the period 2000–2019. These are the exceptions rather than the rule, but with the right policies, regulations and financing for conservation, it is possible to extend these triumphs further and deeper offshore. In December 2020, a panel of international ocean experts, working under the umbrella of the High Level Panel for a Sustainable Ocean Economy, concluded – from three years of research, carried out by 250 authors from 48 nations – that a healthy ocean, 30 per cent protected, could deliver the following: 20 per cent of the carbon emission reductions needed to achieve the Paris Accord's goal to limit warming to 1.5°C (2.7°F); forty times more renewable energy than was generated in 2018; six times more sustainable seafood; 12 million jobs; and US$15.5 trillion in net economic benefits. With a High Seas Treaty agreed, we now stand a chance of actually achieving these goals.

At times, on this journey, I felt an overwhelming sense of gratitude for the ocean, and all that it does for us. Over the last few years, when not travelling, I've been writing from my home in south County Dublin, not far from where I grew up. From here, I can glimpse the sea, and those waters that first enthralled me. Their constant movement reminds me of the vivid, living world beneath. I've been drift-diving here, in Dalkey Sound – a small stretch of water that separates the charming Dalkey Island from the mainland – many times, and in my mind's eye, I can see the kelp beds, lobsters, crabs, scallops, conger eels and fish that gladden the seafloor. When the sun's rays catch the ripples, they shimmer, beckoning me in. Looking out on those waters, I often reflect on how we've come to think of the high seas somehow as other, as separate from the coastal shallows. But they are one and the same, all ocean, which we've managed to deconstruct into various parts that are governed by different nations, under different rules, or by none at all. My foray into the lesser-known world offshore has made me consider how connected we all are to the ocean – unclaimed or otherwise – in almost every aspect of our daily lives. Therein lies an opportunity for change.

ACKNOWLEDGEMENTS

I THINK IT'S FAIR to say that when I first decided to write this book, I had a sense that it could become a long journey, not unlike a transatlantic voyage. I didn't, however, reckon on it being closer to a circumnavigation of the high seas, with unexpected and often violent storms on the crossing. From the outset, I had two champions of this book, friends who not only encouraged and whole-heartedly supported the endeavour, but who seemed genuinely enthralled by the project. One of those was Gaia Vince, who has been a constant champion of this project, and a cheerleader at the sidelines, from start to finish.

My other constant ally was Emma O'Kane, a dear lifelong friend, whose creativity and work ethic touched the lives of everyone who knew her and who watched her perform, whether at Dublin's Abbey Theatre or at Sydney Opera House. Roughly halfway through this project, I lost Emma to a rare and aggressive cancer, caught late because of the Covid pandemic. It was an unbelievable blow, and in spite of her diagnosis, I didn't accept the outcome until she had passed. Nor did I see the Covid pandemic, the cancellation of planned sea trips, and two consecutive years of home schooling my children as being part of this journey. In any case, I was stuck in the doldrums for quite some time before the wind caught my sails again and set me back on course.

There were, as with any project of this size, many who helped me along the way and to whom I am greatly indebted. Thanks to Rupert for holding the fort during my time away from home, and for supporting the project in more ways than are possible to mention. Huge thanks also to George, my son, for his unwavering interest in science, in nature, in the ocean and in octopuses, and to Millie, my daughter, for her insightful advice on handling the trickiest of negotiations. To Marie-Louise, for those first fun adventures to rock pools with Coastwatch Europe, and to Adrian, for sharing with me those unforgettable days in Pelagos.

To Georgina and to my mum, for listening and reading.

To my amazing agent Patrick Walsh, of PEW Literary, a world of thanks – for believing in this project at the outset and supporting it all the way through. To my editors – Ed Lake and Izzy Everington at Profile Books and to Rob Sanders and Nancy Flight at Greystone – for all of your insight, advice and patient guidance, I offer my sincere thanks. Thanks also to Mark Ellingham at Profile Books for invaluable comments, and to Jen Gauthier at Greystone for her support of the project.

I'm extremely grateful to those people and organisations who offered me a place on their research expeditions or who welcomed me into their institutes and organisations, so that I could report this story from the front line. Above all this includes Greenpeace International for their overwhelming generosity and interest in the book, but also Henko de Stigter from NIOZ, IHC, Tethys Research Institute, Global Fishing Watch, SkyTruth, SINTEF, NHM and Callum Roberts, now at the University of Exeter. Thanks also to those who extended offers I was unable to accept owing to the Covid pandemic, including NASA, WHOI and Sea Shepherd.

A huge thanks to those generous few who kindly proofread earlier drafts of this book, including Patrick Walsh of PEW

Literary; Jessica Aldred of *China Dialogue*; Chelsea Noack and Brook Glass-O'Shea at Johns Hopkins; Jennifer Jacquet at New York University; Robert Blasiak of the Stockholm Resilience Centre; Diva Amon of the University of California, Santa Barbara; Rosalind Amon and Annabel Egan.

Crucial to reporting this book were the many experts who gave me their time so generously.

Above all, I owe huge thanks Kristina Gjerde of IUCN for being a willing source on just about every topic in this book and, more broadly, for being a source of energy and inspiration to just about everyone who is interested in the high seas. Huge thanks also to Daniel Kachelriess, also of IUCN, for his support and enthusiasm from start to finish. Specifically, on the topic of fisheries, I'd like to thank Dana Miller of Oceana; Daniel Pauly; Rashid Sumaila; Bjorn Bergman and the team from Global Fishing Watch, including Paul Woods, David Kroodsma and Guillermo Ortuño Crespo; Eduardo Grimaldo and Dag Standal of SINTEF Norway; Martin Weich of the University of Bergen; Webjørn Melle from IMR Bergen; BIM in Ireland and the Marine Institute in Ireland; and Isabel Jarrett at the Pew Trusts.

On deep-sea mining, I would like to thank Diva Amon at the University of California and Adrian Glover of the UK Natural History Museum; Henko de Stigter of NIOZ; Hjalmar Thiel, formerly of the University of Hamburg; Laurens de Jonge of IHC; and many others including Duncan Currie of the High Seas Alliance; Thomas Peacock of MIT; Daniel Jones at the University of Southampton; and Cindy Van Dover of Duke University. On the topic of climate intervention, I'd like to thank Kelly Wanser of SilverLining, Greg Rau of Planetary Technologies, Leslie Field of the Arctic Ice Project, Eric Matzner of Project Vesta and Ulf Riebesell of the University of Kiel, Philip Boyd at the University of Tasmania, among others.

On the topics of marine genetic resources and ocean drug discovery, I would especially like to extend huge thanks to David Newman, formerly of the US National Institutes of Health, for many lengthy and open conversations during the pandemic on the history of marine drug discovery. David, your knowledge was a bottomless well from which I drew, and I'm ever so grateful to you for that time. I would also like to thank Kate Duncan at the University of Strathclyde, and both Marcel Jaspars and Abbe Brown at the University of Aberdeen for many wonderful conversations. Marcel, thank you so much for your time and generosity! Thanks also to Elasmogen; Harriet Harden-Davies of the Deep Ocean Stewardship Initiative; Rachel Wynberg of the University of Cape Town; Peter Girguis at Harvard University; Michael Kanu, deputy permanent representative to the UN for Sierra Leone; Pat Roy Mooney from the Coalition Against Biopiracy; Christian Tiambo at the International Livestock Research Institute in Nairobi; Siva Thambisetty at the London School of Economics; Muriel Rabone at London's Natural History Museum; Shirley Pomponi of Florida Atlantic University; Roger Linington at Simon Fraser University and to Robert Blasiak of the Stockholm Resilience Centre.

On the BBNJ negotiations, the law of the sea and its history, I would like to thank Kristina Gjerde of IUCN, Surabhi Ranganathan at the University of Cambridge and Peggy Kalas of the High Seas Alliance. Thanks also to Vito De Lucia at the Arctic University of Norway in Tromsø for providing insightful thoughts and discussion on the topic of splashdowns and the wider issue of ocean dumping and pollution. Thank you to ESA and to Stijn Lemmens in particular for answering all of my questions on the growing problem of space debris, and on the role of managing space debris. Thanks to Geraint Tarling at the British Antarctic

Survey for sharing his expertise and thoughts on iceberg towing and its potential environmental impacts; on this topic, thanks also to Peter Wadhams, professor emeritus at the University of Cambridge. To Abdulla Alshehi, a sincere thank you for discussing your iceberg harvesting plans. Thank you to Ed Kean of Iceberg Water, to Stephen Bruneau of Memorial University in Newfoundland, and Anita Lam of York University in Canada.

From my time in the Arctic, I'd also like to thank Till Wagner and Hillary Glandon for sharing so much of their expertise with me, as well as for including me in their field work and trusting me with taking their ice cores. Thanks also to Fanny Groundstroem at Greenpeace Nordic for her excellent coordination of the Arctic trip. Also on the Arctic, I'd like to thank Klaus Dodds at Royal Holloway, University of London; Mattias Cape at the University of Washington; Heather Koopman at the University of North Carolina; and Ilka Peeken at the Alfred Wegener Institute in Bremerhaven, Germany.

Similarly, heartfelt thanks to Julia Zanolli for providing communications support, and excellent company, during my time in Antarctica with a research team who could not have been more enjoyable to hang out with. Steve Forrest, Alex, Michael and Noah – it was gutting to leave *Esperanza* and the pleasure of your company. Thanks to the team on *Espi* and on *Arctic Sunrise* for welcoming me more than once, and for sharing your stories with me. Thanks in particular to Laurence for sharing your room on *Arctic Sunrise*; I'm sure it wasn't easy, given your schedule. On Antarctic governance, fisheries and ecology, thanks to Cassandra Brooks at the University of Colorado, not just for a great conversation but for all of your insightful papers, to Heather Lynch at Stony Brook University and to Phil Trathan at the British Antarctic Survey.

On the topic of marine protected areas, I would like to thank Callum Roberts of the University of Exeter and Bethan O'Leary at the University of York for talking me through their Marxan analysis back in 2019, and for all of their invaluable work on this topic. Also Nino Pierantonio for a glorious week in Pelagos back in 2019. Thanks also to Lance Morgan, Beth Pike and the team at the Marine Conservation Institute for creating MPAtlas, for answering all of my questions on high seas MPAS. On the topic of sperm whales, sincere thanks to Nino Pierantonio of Tethys and also to Taylor Hersh at Dalhousie University.

Thanks also to Will McCallum, Luke Massey, John Sauven, Lucie Marks and Julia Zanolli of Greenpeace for your practical support of this project. Thank you to Johns Hopkins University and in particular to my colleague Melissa Hendricks for passing on such great contacts. Thanks also to those editors at various magazines who commissioned me at early, crucial stages of my reporting, including Rob Kunzig at *National Geographic*, Mark Fischetti at *Scientific American*, Richard Van Noorden at *Nature*, Jessica Aldred at *China Dialogue*, Sarah McPherson at *BBC Wildlife* and João Medeiros at *WIRED*.

Thanks also to Dana Miller of Oceana for answering so many of my early questions on fisheries and pirate fishing, and to Mike Walker, who works with the Antarctic and Southern Ocean Coalition, for his knowledge, but even more for his enduring optimism on the possibilities for ocean protection.

There are many others who, through sharing their time and expertise, helped me to report this book, including Marcus and Brent at Hitra Fish Farm by guiding me through the world of Norwegian fish farming; Carlos Duarte at KAUST; Xabier Irigoien at AZTI and Adrian Martin of the University of Southampton on mesopelagic fishing; Conor Nolan, Emmet Jackson and Michael Gallagher at BIM Ireland on mesopelagics; also Peter Heffernan and David Reid at the Marine Institute Ireland. Heartfelt thanks to Julia Blanchard at the University of Tasmania and Audrey Darnaude of the University of Montpellier for half a lifetime of great ocean-themes conversations! I hope we can continue them for many years to come. There are many others whose help I've likely overlooked here – so apologies if I've not mentioned you – but know that your time and effort have been appreciated and have helped hugely in bringing these ideas to life.

REFERENCES

CHAPTER 1: THE OUTER SEA

T.A. Branch, 'A review of orange roughy *Hoplostethus atlanticus* fisheries, estimation methods, biology and stock structure', *South African Journal of Marine Science* 23, no. 1 (8 April 2010): doi.org/10.2989/025776101784529006.

D. Cordingly, *Under the Black Flag: The Romance and the Reality of Life Among the Pirates* (Random House, 1996).

R. Corfield, *The Silent Landscape: The Scientific Voyage of HMS Challenger* (Joseph Henry Press, 2003).

The Economist, 'In deep water', 24 February 2014, economist.com/international/2014/02/24/in-deep-water.

H. Grotius, *The Free Sea*, translated by R. Hakluyt with W. Welwod's critique and Grotius's reply, edited and with an introduction by D. Armitage (Liberty Find, 2004): scholar.harvard.edu/files/armitage/files/free_sea_ebook.pdf.

B.S. Halpern et al., 'Recent pace of change in human impact on the world's ocean', *Scientific Reports* 9, art. 11609 (2019): doi.org/10.1038/s41598-019-47201-9.

M.E. Hoare (ed.), *The Resolution Journal of Johann Reinhold Forster, 1772–1775*, vol. IV (Routledge, 1981).

J.-B. Jouffray et al., 'The Blue Acceleration: The trajectory of human expansion into the ocean', *One Earth* 2, no. 1 (24 January 2020): doi.org/10.1016/j.oneear.2019.12.016.

J. Kämpf and P. Chapman, *Upwelling Systems of the World: A Scientific Journey to the Most Productive Marine Ecosystems* (Springer, 2016).

H. Nellen (ed.), *Hugo Grotius: A Lifelong Struggle for Peace in Church and State, 1583–1645* (Brill, 2015).

J. Nigg, *Sea Monsters: A Voyage Around the World's Most Beguiling Map* (University of Chicago Press, 2013).

L. Paine, *The Sea & Civilization: A Maritime History of the World* (Atlantic Books, 2014).

R. Picheta, 'CO2 levels in the atmosphere reach a 3 million-year high, putting the world "way off track" on climate goals', CNN, 25 October 2021, edition.cnn.com/2021/10/25/world/emissions-climate-greenhouse-gas-bulletin-wmo-intl/index.html.

S. Rahmstorf, 'Thermohaline ocean circulation', in *Encyclopedia of Quaternary Sciences*, ed. S.A. Elias (Elsevier, 2006).

E. Ramirez-Llodra et al., 'Man and the last great wilderness: Human impact on the deep sea', *PLOS ONE* 6, no. 7, art. e22588 (2011): doi.org/10.1371/journal.pone.0022588.

C.M. Roberts, 'Deep impact: The rising toll of fishing in the deep sea', *Trends in Ecology & Evolution* 17, no. 5 (2002): doi.org/10.1016/S0169-5347(02)02492-8.

D. Rothwell (ed.) et al., *The Oxford Handbook of the Law of the Sea* (Oxford University Press, 2015).

H.M. Rozwadowski, *Fathoming the Ocean: The Discovery and Exploration of the Deep Sea* (Belknap, 2008).

H.M. Rozwadowski, *Vast Expanses: A History of the Oceans* (Reaktion Books, 2018).

B. Specktor, 'Project to map entire ocean floor by 2030 passes 20% mark', *Live Science*, 23 June 2001, livescience.com/map-20-percent-global-seabed.html.

S.A. Swarztrauber, 'The three-mile limit of territorial seas: A brief history', PhD diss., Faculty of the School of International Service, American University (1970): hdl.handle.net/10945/15200.

30×30: A Blueprint for Ocean Protection (University of Oxford, University of York and Greenpeace, 2019): greenpeace.org/30x30.

M.J. van Ittersum, 'Hugo Grotius in context: Van Heemskerck's capture of the *Santa Catarina* and its justification in *de jure praedae* (1604–1606)', *Asian Journal of Social Science* 31, no. 3 (2003): doi.org/10.1163/156853103322895360.

L. Victorero et al., 'Out of sight, but within reach: A global history of bottom-trawled deep-sea fisheries from >400 m depth', *Frontiers in Marine Science* (11 April 2018): doi.org/10.3389/fmars.2018.00098.

WHO, 'High levels of antibiotic resistance found worldwide, new data shows', 29 January 2018, who.int/news/item/29-01-2018-high-levels-of-antibiotic-resistance-found-worldwide-new-data-shows.

CHAPTER 2: ENTER THE TWILIGHT ZONE

R.O. Amoroso et al., 'Comment on "Tracking the global footprint of fisheries"', *Science* 361, no. 6404 (24 August 2018): doi.org/10.1126/science.aat671.

K. Buesseler, 'The Ocean Twilight Zone's role in climate change', Ocean Twilight Zone, Woods Hole Oceanographic Institution, 16 February 2022: doi.org/10.1575/1912/28074.

K.O. Buesseler et al., 'Metrics that matter for assessing the ocean biological carbon pump', *Proceedings of the National Academy of Sciences* 117, no. 18 (2020): doi.org/10.1073/pnas.1918114117.

K.O. Buesseler et al., 'Revisiting carbon flux through the ocean's twilight zone', *Science* 316, no. 5824 (27 April 2007): doi.org/10.1126/science.113795.

M.G. Burgess et al., 'Protecting marine mammals, turtles, and birds by rebuilding global fisheries', *Science* 359, no. 6381 (16 March 2018): doi.org/10.1126/science.aao4248.

M. Drexler, 'The Ocean Twilight Zone's crucial carbon pump', Woods Hole Oceanographic Institution, 9 January 2020, whoi.edu/news-insights/content/the-ocean-twilight-zones-crucial-carbon-pump/.

C.M. Duarte, 'Seafaring in the 21st century: The Malaspina 2010 circumnavigation expedition', *Bulletin of Limnology and Oceanography* 24, no. 1 (February 2015): doi.org/10.1002/lob.10008.

Federation of Norwegian Industries, 'Summary: Roadmap for the aquaculture industry – sustainable growth', 25 July 2017, norskindustri.no/siteassets/dokumenter/rapporter-og-brosjyrer/veikart-for-havbruksnaringen---kort-versjon_eng.pdf.

The Fishing Daily, 'EU fishermen strongly rejects Norway's unilateral mackerel quota grab', 28 May 2021, thefishingdaily.com/featured-news/eu-fishermen-strongly-rejects-norways-unilateral-mackerel-quota-grab/.

D. Freestone et al., 'World Heritage in the high seas: an idea whose time has come', UNESCO and International Union for Conservation of Nature (2016): unesdoc.unesco.org/ark:/48223/pf0000245467.

K.M. Gjerde, G. Wright and C. Durussel, 'Strengthening high seas governance through enhanced environmental assessment processes: A case study of mesopelagic fisheries and options for a future BBNJ treaty', STRONG High Seas Project (2021): doi.org/10.48440/iass.2021.001.

J. Gjøsaeter and K. Kawaguchi, *A Review of the World Resources of Mesopelagic Fish*, FAO Fisheries Technical Paper No. 193 (FAO, 1980).

E. Grimaldo et al., 'Investigating the potential for a commercial fishery in the Northeast Atlantic utilizing mesopelagic species', *ICES Journal of Marine Science* 77, nos. 7–8 (December 2020): doi.org/10.1093/icesjms/fsaa114.

B. Haas et al., 'Factors influencing the performance of regional fisheries management organizations', *Marine Policy* 113, art. 103787 (2020): doi.org/10.1016/j.marpol.2019.103787.

International Union for Conservation of Nature, 'High seas gems: hidden treasures of our blue earth', 23 October 2008, iucn.org/content/high-seas-gems-hidden-treasures-our-blue-earth.

X. Irigoien et al., 'Large mesopelagic fishes biomass and trophic efficiency in the open ocean', *Nature Communications* 5, art. 3271 (2014): doi.org/10.1038/ncomms4271.

D. Kroodsma et al., 'Tracking the global footprint of fisheries', *Science* 359, no. 6378 (23 February 2018): doi.org/10.1126/science.aao564.

V. Lam and D. Pauly, 'Mapping the global biomass of mesopelagic fishes', *Sea Around Us Project Newsletter*, July/August 2005.

K. McVeigh, 'Stop overfishing or we'll buy elsewhere, top UK fish firm warns European states', *Guardian*, 22 October 2021, theguardian.com/environment/2021/oct/22/stop-overfishing-or-well-buy-elsewhere-top-uk-fish-firm-warns-european-states.

'Mesopelagic initiative: Unleashing new resources for a growing human population', Norwegian Institute of Marine Research, University of Bergen (2017): semanticscholar.org/paper/Mesopelagic-Initiative%3A-Unleashing-new-marine-for-a/07a56230d32b728933d8b57a80d8099ca06c2794.

Mission Blue Hope Spots: mission-blue.org/hope-spots/.

Norwegian Seafood Council: norwegianseafoodcouncil.com.

Ocean Twilight Zone, Woods Hole Oceanographic Institution: twilightzone.whoi.edu.

N. Pacoureau et al., 'Half a century of global decline in oceanic sharks and rays', *Nature* 589 (2021): doi.org/10.1038/s41586-020-03173-9.

M.T. Petersson, 'Transparency in global fisheries governance: The role of non-governmental organizations', *Marine Policy* 136, art. 104128 (2022): doi.org/10.1016/j.marpol.2020.104128.

Pew Charitable Trusts, 'Underwater treasures of the high seas', March 2016, pewtrusts.org/highseas.

R. Rayfuse, 'Section 20, regional fisheries management organisations', in *The Oxford Handbook of the Law of the Sea*, ed. D. Rothwell et al. (Oxford University Press, 2015).

Research Council of Norway, 'Harvesting the mesopelagics – ecological and management implications' (2018–21): prosjektbanken.forskningsradet.no/en/project/FORISS/280546?Kilde=FORISS&distribution=Ar&chart=bar&calcType=funding&Sprak=no&sortBy=date&sortOrder=desc&resultCount=30&offset=90&Prosjektleder=Vidar+Kolstad; Working Group on Zooplankton Ecology, prep.ices.dk/community/Documents/Science%20EG%20ToRs/EPDSG/2018/WGZE%20MA%20ToRs%202017.pdf.

C. Roberts, *The Unnatural History of the Sea*, 6th ed. (Island Press, 2007).

H.M. Rozwadowski, *Vast Expanses: A History of the Oceans* (Reaktion Books, 2018).

E. Sala et al., 'The economics of fishing the high seas', *Science Advances* 4, no. 6 (6 June 2018): doi.org/10.1126/sciadv.aat2504.

L. Schiller et al., 'High seas fisheries play a negligible role in addressing global food security', *Science Advances* 4, no. 8 (8 August 2018): doi.org/10.1126/sciadv.aat8351.

Seafood From Norway: fromnorway.com/seafood-from-norway/salmon/.

M. Silver, 'Marine snow: A brief historical sketch', *Bulletin Limnology and Oceanography* 24, no. 1 (February 2015): doi.org/10.1002/lob.10005.

D. Standal, 'The rise and decline of blue whiting fisheries – capacity expansion and future regulations', *Marine Policy* 30, no. 4 (2006): doi.org/10.1016/j.marpol.2005.03.007.

D. Standal and E. Grimaldo, 'Institutional nuts and bolts for a mesopelagic fishery in Norway', *Marine Policy* 119, art. 104043 (September 2020): doi.org/10.1016/j.marpol.2020.104043.

D. Standal and E. Grimaldo, 'Lost in translation? Practical and scientific input to the mesopelagic fisheries discourse', *Marine Policy* 134, art. 104785 (December 2021): doi.org/10.1016/j.marpol.2021.104785.

U.R. Sumaila et al., 'Updated estimates and analysis of global fisheries subsidies', *Marine Policy* 109, art. 103695 (2019): doi.org/10.1016/j.marpol.2019.103695.

W. Swartz et al., 'The spatial expansion and ecological footprint of fisheries (1950 to present)', *PLOS ONE* 5, no. 12, art. e15143 (2010): doi.org/10.1371/journal.pone.0015143.

30×30: A Blueprint for Ocean Protection (University of Oxford, University of York and Greenpeace, 2019): greenpeace.org/30x30.

Vertical Migration, Superflex Studio, superflex.net/works/vertical_migration; see also youtube.com/watch?v=vxw3GdjDzsQ.

R.A. Watson and T. Morato, 'Fishing down the deep: Accounting for within-species changes in depth of fishing', *Fisheries Research* 140 (February 2013): doi.org/10.1016/j.fishres.2012.12.004.

L. Woodall et al., 'Function of the high seas and anthropogenic impacts science update 2012–2017', University of Oxford for the High Seas Alliance, July 2017, highseasalliance.org/wp-content/uploads/2017/07/HS-Synthesis-Oxford-110717.pdf.

D. Yutian, 'Pacific bluefin quota increase was too hasty', *China Dialogue Ocean*, 29 March 2022, chinadialogueocean.net/en/fisheries/pacific-bluefin-quota-increase-was-too-hasty/.

CHAPTER 3: THE HUNT FOR DARK TARGETS

A.I. Arkhipkin et al., 'World squid fisheries', *Reviews in Fisheries Science & Aquaculture* 23, no. 2 (2015): doi.org/10.1080/23308249.2015.1026226.

B. Bergman, 'Tracking the Chinese squid fleet in the South Pacific – part 1: Voyage to the Galapagos', SkyTruth, 30 October 2018, skytruth.org/2018/10/tracking-the-chinese-squid-fleet-in-the-south-pacific-part-1/.

S. Bladen, 'The capture of the MV *NIKA*: A case of illicit fishing and a showcase for how to beat it', *Global Fishing Watch*, 23 July 2019, globalfishingwatch.org/transparency/the-capture-of-the-mv-nika-a-case-of-illicit-fishing-and-a-showcase-for-how-to-beat-it/.

Blue Justice, FishCRIME: bluejustice.org/symposium/fishcrime-2016/.

E. Bonaccorso et al., 'International fisheries threaten globally endangered sharks in the Eastern Tropical Pacific Ocean: The case of the *Fu Yuan Yu*

Leng 999 reefer vessel seized within the Galápagos Marine Reserve', *Scientific Reports* 11, art. 14959 (2021): doi.org/10.1038/s41598-021-94126-3.

Z. Chun, 'China announces closed season on squid spawning grounds', *China Dialogue Ocean*, 18 June 2020, chinadialogueocean.net/en/fisheries/14146-china-announces-closed-season-squid-spawning-grounds/.

Z. Chun, 'China targets distant-water criminals with new fisheries law', *China Dialogue Ocean*, 21 January 2020, chinadialogueocean.net/en/fisheries/12714-china-fisheries-law-distant-water-fishing/.

D. Collyns, '"They just pull up everything!" Chinese fleet raises fears for Galápagos sea life', *Guardian*, 6 August 2020, theguardian.com/environment/2020/aug/06/chinese-fleet-fishing-galapagos-islands-environment.

E. Engdal and K. Sæter, *Catching Thunder: The True Story of the World's Longest Sea Chase* (Zed Books, 2018).

FAO, *The State of World Fisheries and Aquaculture 2022: Towards Blue Transformation*, fao.org/3/cc0461en/cc0461en.pdf.

Global Fishing Watch: globalfishingwatch.org.

Greenpeace International, 'Squids in the spotlight: Unregulated squid fisheries are headed for disaster', March 2022, greenpeace.org.uk/wp-content/uploads/2022/03/e13337d8-squids-in-the-spotlight.pdf.

O. Heffernan, 'The hidden fight to stop illegal fishing from destroying our oceans', WIRED, 19 March 2019, wired.co.uk/article/illegal-fishing-global-fishing-watch.

R. McDowell, M. Mason and M. Mendoza, 'AP investigation: Slaves may have caught the fish you bought', Associated Press, 25 March 2015, ap.org/explore/seafood-from-slaves/ap-investigation-slaves-may-have-caught-the-fish-you-bought.html.

H. Moustahfid et al., 'Ecological-fishery forecasting of squid stock dynamics under climate variability and change: Review, challenges, and recommendations', *Reviews in Fisheries Science & Aquaculture* 29 no. 4 (2020): doi.org/10.1080/23308249.2020.1864720.

North Atlantic Fisheries Intelligence Group and INTERPOL, *Chasing Red Herrings: Flags of Convenience and the Impact on Fisheries Crime Law Enforcement* (NA-FIG: 2017).

F. Obermaier and B. Obermayer, 'The downfall of a scandalous firm', Panama Papers: The Secrets of Dirty Money, trans. Valerié Callaghan, *Süddeutsche Zeitung*, April 2017, panamapapers.sueddeutsche.de/articles/e344090/.

D. Pauly and D. Zeller, 'Catch reconstructions reveal that global marine fisheries catches are higher than reported and declining', *Nature Communications* 7, art. 10244 (2016): doi.org/10.1038/ncomms10244.

J. Ruiz, I. Caballero and G. Navarro, 'Sensing the same fishing fleet with AIS and VIIRS: A seven-year assessment of squid jiggers in FAO Major Fishing Area 41', *Remote Sensing* 12, no. 1 (2020): doi.org/10.3390/rs12010032.

W.H.H. Sauer et al., 'World octopus fisheries', *Reviews in Fisheries Science & Aquaculture* 29, no. 3 (2019): doi.org/10.1080/23308249.2019.1680603.

M. Valentine, 'Oceana finds 300 Chinese vessels pillaging the Galapagos for squid', *Zenodo*, 16 September 2020, doi.org/10.5281/zenodo.4118526.

CHAPTER 4: TREASURES FROM THE DEEP

M. Allen, 'An intellectual history of the common heritage of mankind as applied to the oceans', master's thesis, Department of Marine Affairs, University of Rhode Island, paper 1088 (1992): digitalcommons.uri.edu/theses/1088.

D.J. Amon et al., 'Insights into the abundance and diversity of abyssal megafauna in a polymetallic-nodule region in the eastern Clarion-Clipperton Zone', *Scientific Reports* 6, art. 30492 (2016): doi.org/10.1038/srep30492.

Blue Nodules: blue-nodules.eu/#:~:text=blue%20nodules%20is%20a%20research,floor%20with%20minimum%20environmental%20impact.

J. Buttigieg, 'Arvid Pardo – a diplomat with a mission', *Symposia Melitensia* 12 (2016): core.ac.uk/download/pdf/83020273.pdf.

R. Corfield, *The Silent Landscape: The Scientific Voyage of HMS Challenger* (Joseph Henry Press, 2003).

A. Davies and B. Doherty, 'Corruption, incompetence and a musical: Nauru's cursed history', *Guardian*, 3 September 2018, theguardian.com/world/2018/sep/04/corruption-incompetence-and-a-musical-naurus-riches-to-rags-tale.

B. Doherty, 'Collapse of PNG deep-sea mining venture sparks calls for moratorium', *Guardian*, 15 September 2019, theguardian.com/world/2019/sep/16/collapse-of-png-deep-sea-mining-venture-sparks-calls-for-moratorium.

J.C. Drazen et al., 'Midwater ecosystems must be considered when evaluating environmental risks of deep-sea mining', *Proceedings of the National Academy of Sciences* 117, no. 30 (8 July 2020): doi.org/10.1073/pnas.2011914117.

DSM Observer, 'GSR Patania II suffers a setback', 21 March 2019, dsmobserver .com/2019/03/GSR-PATANIA-II-SUFFERS-A-SET-BACK/.

I. Feichtner, 'Sharing the riches of the sea: The redistributive and fiscal dimension of deep seabed exploitation', *European Journal of International Law* 30, no. 2 (May 2019): doi.org/10.1093/ejil/chz022.

E.J. Foell, H. Thiel and G. Schriever, 'DISCOL: A long-term, large-scale, disturbance-recolonization experiment in the abyssal eastern tropical South Pacific Ocean', paper presented at the Offshore Technology Conference, Houston, TX (May 1990): doi.org/10.4043/6328-ms.

A. Garanovic, 'Deep sea mining robot gets stuck 4,500 metres beneath Pacific Ocean's surface', *Offshore Energy*, 29 April 2021, offshore-energy.biz/deep-sea-mining-robot-gets-stuck-4500-metres-beneath-pacific-oceans-surface/.

O. Heffernan, 'Seabed mining is coming – bringing mineral riches and fears of epic extinctions', *Nature* 571, no. 7766 (July 2019): doi.org/10.1038/d41586-019-02242-y.

K. Hund et al., *Minerals for Climate Action: The Mineral Intensity of the Clean Energy Transition*, Climate-Smart Mining Facility and World Bank Group, 2020, pubdocs.worldbank.org/en/961711588875536384/Minerals-for-Climate-Action-The-Mineral-Intensity-of-the-Clean-Energy-Transition.pdf.

International Seabed Authority Mining Code: isa.org.jm/the-mining-code/draft-exploitation-regulations/.

D. Johnson and M.A. Ferreira, 'ISA areas of particular environmental interest in the Clarion-Clipperton Fracture Zone', *International Journal of Marine and Coastal Law* 30, no. 3 (2015): doi.org/10.1163/15718085-12341367.

D.O.B. Jones et al., 'Biological responses to disturbance from simulated deep-sea polymetallic nodule mining', *PLOS ONE* 12, no. 2, art. e0171750 (2017): doi.org/10.1371/journal.pone.0171750.

D.O.B. Jones et al., 'Environment, ecology, and potential effectiveness of an area protected from deep-sea mining (Clarion Clipperton Zone, abyssal Pacific)', *Progress in Oceanography* 197, art.102653 (2021): doi.org/10.1016/j.pocean.2021.102653.

R. Kirchain, F.R. Field and R. Roth, 'Financial regimes for polymetallic nodule mining: A comparison of four economic models', Materials Systems Laboratory, MIT, 2019.

M. Lodge et al., 'Seabed mining: International Seabed Authority environmental management plan for the Clarion–Clipperton Zone. A partnership approach', *Marine Policy* 49 (2014): doi.org/10.1016/j.marpol.2014.04.006.

S. Manoochehri et al., 'A contribution to future critical raw materials recycling CEWASTE project final report', CEWASTE, 2021, cewaste.eu/wp-content/uploads/2021/04/CEWASTE-Final-Public-Raport.pdf.

N. Niarchos, 'The dark side of Congo's cobalt rush', *New Yorker*, 31 May 2021, newyorker.com/magazine/2021/05/31/the-dark-side-of-congos-cobalt-rush.

H.J. Niner et al., 'Deep-sea mining with no net loss of biodiversity – an impossible aim', *Frontiers in Marine Science* 5 (2018): doi.org/10.3389/fmars.2018.00053.

Ocean Mining Intel, 'Main shareholders gain control of Nautilus Minerals', 22 August 2019, oceanminingintel.com/news/industry/main-shareholders-gain-control-of-nautilus-minerals.

B.N. Orcutt et al., 'Impacts of deep-sea mining on microbial ecosystem services', *Limnology and Oceanography* 65, no. 7 (July 2020): doi.org/10.1002/lno.11403.

T. Peacock and M. Alford, 'The race is on to mine and protect the deep seafloor', *Scientific American*, 1 May 2018, doi.org/10.1038/scientificamerican0518-72.

H. Reid, 'Renault and U.S. carmaker Rivian back moratorium on deep-sea mining' Reuters, 11 February 2022, tinyurl.com/2p99z2uz.

E. Simon-Lledó et al., 'Biological effects 26 years after simulated deep-sea mining', *Scientific Reports* 9, art. 8040 (2019): doi.org/10.1038/s41598-019-44492-w.

H. Thiel et al., 'The large-scale environmental impact experiment DISCOL – reflection and foresight', *Deep Sea Research Part II: Topical Studies in Oceanography* 48, nos. 17–18 (2001): doi.org/10.1016/s0967-0645(01)00071-6.

N. Toro, P. Robles and R.I. Jeldres, 'Seabed mineral resources, an alternative for the future of renewable energy: A critical review', *Ore Geology Reviews* 126, art. 103699 (2020): doi.org/10.1016/j.oregeorev.2020.103699.

M. Trancossi, 'An overview of scientific and technical literature on Coanda effect applied to nozzles', SAE *Technical Papers* no. 2591 (2011): doi.org/10.4271/2011-01-2591.

R. Turner, 'Reimagining rare earth elements in a sacrifice zone-free future', GreenBiz, (6 February 2019), greenbiz.com/article/reimagining-rare-earth-elements-sacrifice-zone-free-future.

T.W. Washburn et al., 'Ecological risk assessment for deep-sea mining', *Ocean & Coastal Management* 176 (15 June 2019): doi.org/10.1016/j.ocecoaman.2019.04.014.

World Ocean Review, 'Mineral resources: Manganese nodule treasures', 2014, worldoceanreview.com/en/wor-3/mineral-resources/manganese-nodules/#:~:text=It%20is%20fascinating%20how%20extremely,between%2010%20and%20100%20millimetres.

World Resources Institute, 'Ok Tedi mine: Unearthing controversy (Papua New Guinea)', in *World Resources 2002–2004: Decisions for the Earth – Balance, Voice, and Power* (2003), pdf.wri.org/wr2002_case_oktedi_papua.pdf.

'Write rules for deep-sea mining before it's too late' *Nature* 571, no. 447 (2019): doi.org/10.1038/d41586-019-02276-2.

CHAPTER 5: THE INTERVENTIONISTS

M. Allen et al., 'The Oxford principles for net zero aligned carbon offsetting', Smith School of Enterprise and the Environment, University of Oxford, September 2020, smithschool.ox.ac.uk/sites/default/files/2022-01/Oxford-Offsetting-Principles-2020.pdf.

Aspen Institute, 'Guidance for ocean-based carbon dioxide removal: A pathway to developing a code of conduct', 8 December 2021, aspeninstitute.org/publications/ocean-carbon-dioxide-removal/.

V.N. Aswathy et al., 'Climate extremes in multi-model simulations of stratospheric aerosol and marine cloud brightening climate engineering', *Atmospheric Chemistry and Physics* 15, no. 16 (2015): doi.org/10.5194/acp-15-9593-2015.

D. Biello, 'Pacific Ocean hacker speaks out', *Scientific American*, 24 October 2012, scientificamerican.com/article/questions-and-answers-with-rogue-geoengineer-carbon-entrepreneur-russ-george/.

K. Bower et al., 'Amelioration of global warming by controlled enhancement of the albedo and longevity of low-level maritime clouds', paper presented at the 14th International Conference on Clouds and Precipitation, Bologna, Italy (17 July 2004): n2t.net/ark:/85065/d71j98xs.

P.W. Boyd and C.M.G. Vivian, 'High level review of a wide range of proposed marine geoengineering techniques', GESAMP Joint Group of Experts on the Scientific Aspects of Marine Environmental Protection, GESAMP *Reports and Studies* 98 (2019): doi.org/10.25607/OBP-1944.

CBC News, 'B.C. company responds to ocean fertilization lawsuit', 26 February 2014, cbc.ca/news/canada/british-columbia/b-c-company-responds-to-ocean-fertilization-lawsuit-1.2552119.

G. Cooper et al., 'Preliminary results for salt aerosol production intended for marine cloud brightening, using effervescent spray atomization', *Philosophical Transactions of the Royal Society A* 372, no. 2031 (28 December 2014): doi.org/10.1098/rsta.2014.0055.

C. Fake (host), 'Cloud brightening for climate fever', interview with K. Wanser et al. on *Should This Exist?* (podcast), December 2020, shouldthisexist.com/cloud-brightening-for-climate-fever/.

L. Field and A. Strawa, 'Restoring arctic ice: A new way to stabilize the climate', *Arctic Circle Journal*, 9 March 2021, arcticcircle.org/journal/restoring-arctic-ice-a-new-way-to-stabilize-the-climate.

L. Field et al., 'Increasing arctic sea ice albedo using localized reversible geoengineering', *Earth's Future* 6, no. 6, (2018): doi.org/10.1029/2018EF000820.

J.R. Fleming, *Fixing the Sky: The Checkered History of Weather and Climate Control* (Columbia University Press, 2012).

M. Harris, 'Billionaire geoengineering', *New Scientist*, September 2020, newscientist.com/article/mg24732980-500-silicon-valley-billionaires-want-to-geoengineer-the-worlds-oceans.

R. Hamblyn and the Met Office, *The Met Office Cloud Book – Updated Edition: How to Understand the Skies* (David & Charles, 2021).

O. Heffernan, 'Who rules climate intervention on the high seas?' *China Dialogue*, 27 August 2020, chinadialogue.net/en/climate/climate-intervention-on-the-high-seas/.

A.-M. Hubert, 'A code of conduct for responsible geoengineering research', *Global Policy* 12, no. S1 (April 2021): doi.org/10.1111/1758-5899.12845.

IPCC, 'Summary for policymakers', in *Global Warming of 1.5 °C: An IPCC Special Report on the Impacts of Global Warming of 1.5 °C Above Pre-Industrial Levels and Related Global Greenhouse Gas Emission Pathways, in the Context of Strengthening the Global Response to the Threat of Climate Change, Sustainable*

Development, and Efforts to Eradicate Poverty, ed. Masson-Delmotte et al. (Cambridge University Press, 2018), doi.org/10.1017/9781009157940.001.

O. Khazan, 'The brain bro', *Atlantic*, October 2016, theatlantic.com/magazine/archive/2016/10/the-brain-bro/497546/.

J. Latham, 'Control of global warming?' *Nature* 347 (1990): doi.org/10.1038/347339b0.

National Academies of Sciences, Engineering, and Medicine, *A Research Strategy for Ocean-Based Carbon Dioxide Removal and Sequestration* (National Academies Press, 2022): doi.org/10.17226/26278.

K. Piper, 'The climate renegade', *Vox*, 4 June 2019, vox.com/the-highlight/2019/5/24/18273198/climate-change-russ-george-unilateral-geoengineering.

Project Vesta: vesta.earth.

S. Salter, G. Sortino and J. Latham, 'Sea-going hardware for the cloud albedo method of reversing global warming', *Philosophical Transactions of the Royal Society A* 266, no. 1882 (2008): doi.org/10.1098/rsta.2008.0136.

V. Smetacek et al., 'Deep carbon export from a Southern Ocean iron-fertilized diatom bloom', *Nature* 487 (2012): doi.org/10.1038/nature11229.

C.W. Stjern et al., 'Response to marine cloud brightening in a multi-model ensemble', *Atmospheric Chemistry and Physics* 18, no. 2 (2018): doi.org/10.5194/acp-18-621-2018.

A. Vaughan, 'Scientists want to restore the oceans with artificial whale poo, *New Scientist*, 22 February 2022, newscientist.com/article/2309262-scientists-want-to-restore-the-oceans-with-artificial-whale-poo/#ixzz7sm-j9gy9K.

N.M. Weyer (ed.), 'IPCC, Annex I: Glossary', in *The Ocean and Cryosphere in a Changing Climate*, ed. H.O. Pörtner et al., Special Report of the Intergovernmental Panel on Climate Change (Cambridge University Press, 2019), doi.org/10.1017/9781009157964.010.

XPRIZE Carbon Removal, Musk Foundation: xprize.org/prizes/elonmusk.

S. Zornetzer, A. Strawa and T. Player, 'Restoring arctic ice: A more benign climate intervention?' Arctic Ice Project white paper, April 2021, arcticiceproject.org/wp-content/uploads/2021/05/aip_white paper_04_09_2021_3.pdf.

CHAPTER 6: A 'NEAR-ARCTIC' STATE

M.M. Bennett et al., 'The opening of the Transpolar Sea Route: Logistical, geopolitical, environmental, and socioeconomic impacts', *Marine Policy* 121, art. 104178 (2020): doi.org/10.1016/j.marpol.2020.104178.

Z. Chun, 'China's "arctic silk-road" projects', *China Dialogue*, 9 January 2020, chinadialogueocean.net/en/climate/12569-chinas-arctic-silk-road-projects/.

M. Decamps, 'The ice silk road: Is China a "near-arctic-state"?' Institute for Security & Development Policy, February 2019, isdp.eu/publication/the-ice-silk-road-is-china-a-near-artic-state/.

M. Fischetti, 'Nations claim large overlapping sections of arctic seafloor', *Scientific American*, 1 August 2019, scientificamerican.com/article/nations-claim-large-overlapping-sections-of-arctic-seafloor/.

M. Fischetti, 'Special report: Future of the arctic', *Scientific American*, 1 August 2019, scientificamerican.com/article/special-report-future-of-the-arctic/.

M-V. Guarino et al., 'Sea-ice-free arctic during the Last Interglacial supports fast future loss', *Nature Climate Change* 10 (2020): doi.org/10.1038/s41558-020-0865-2.

R. Halliburton and K. Dodds, 'The battle for the arctic', *Prospect*, May 2021, prospectmagazine.co.uk/world/37480/the-battle-for-the-arctic.

O. Heffernan, 'Melting ice may be a boon for some Arctic whales – then a bust', *National Geographic*, 2 July 2019, nationalgeographic.co.uk/environment-and-conservation/2019/07/melting-ice-may-be-a-boon-for-some-arctic-whales-then-a-bust.

M. Henriques, 'The rush to claim an undersea mountain range', BBC Future, 23 July 2020, bbc.com/future/article/20200722-the-rush-to-claim-an-undersea-mountain-range.

M. Humpert and A. Raspotnik, 'The future of arctic shipping along the Transpolar Sea Route', in *Arctic Yearbook 2012*, ed. Lassi Heininen (Thematic Network on Geopolitics and Security of the University of the Arctic, 2012), arcticyearbook.com/images/yearbook/2012/Scholarly_Papers/14.Humpert_and_Raspotnik.pdf.

International Maritime Organization, International Code for Ships Operating in Polar Waters (Polar Code), 1 January 2017: wwwcdn.imo.org/localresources/en/MediaCentre/HotTopics/Documents/POLAR%20CODE%20TEXT%20AS%20ADOPTED.pdf.

J. Knies et al., 'The emergence of modern sea ice cover in the Arctic Ocean', *Nature Communications* 5, art. 5608 (2014): doi.org/10.1038/ncomms6608.

C. Knudsen et al., 'Samples from the Lomonosov Ridge place new constraints on the geological evolution of the Arctic Ocean', in *Circum-Arctic Lithosphere Evolution*, ed. V. Pease and B. Coakley (Geological Society, London, Special Publications 460, 2018), doi.org/10.1144/SP460.17.

T. Krumpen et al., 'Arctic warming interrupts the Transpolar Drift and affects long-range transport of sea ice and ice-rafted matter', *Scientific Reports* 9, art. 5459 (2019): doi.org/10.1038/s41598-019-41456-y.

R. Kwok, 'Arctic sea ice thickness, volume, and multiyear ice coverage: losses and coupled variability (1958–2018)', *Environmental Research Letters* 13, no. 10 (2018): doi.org/10.1088/1748-9326/aae3ec.

R. Lindsey and M. Scott, 'Climate change: Arctic sea ice summer minimum', NOAA Climate, 18 October 2022, climate.gov/news-features/understanding-climate/climate-change-arctic-sea-ice-summer-minimum.

Maritime Executive, 'China advances arctic ambitions with icebreaker *Xuelong 2*', 12 August 2021, maritime-executive.com/article/china-advances-arctic-ambitions-with-icebreaker-xuelong-2.

F. Nansen, *Farthest North: The Incredible Three-Year Voyage to the Frozen Latitudes of the North* (Modern Library, 1999).

T. Nilsen, '100 consecutive months with above normal temperatures at Svalbard', *Barents Observer*, 25 March 2019, thebarentsobserver.com/en/ecology/2019/03/march-coming-end-svalbard-can-look-back-100-months-above-normal-temperatures#:~:text=With%20March%20coming%20to%20an,with%20the%20Norwegian%20Meteorological%20Institute.

J.E. Overland et al., 'Arctic report card: Surface air temperature' NOAA Arctic, 2019, arctic.noaa.gov/report-card/report-card-2019/surface-air-temperature-4/.

K. Peek and M. Fischetti, 'The great ocean divide', *Scientific American*, 1 August 2019, scientificamerican.com/article/the-great-ocean-divide/.

A. Randelhoff et al., 'Seasonality of the physical and biogeochemical hydrography in the inflow to the Arctic Ocean through Fram Strait', *Frontiers in Marine Science* 5 (29 June 2018): doi.org/10.3389/fmars.2018.00224.

Royal Museums Greenwich, 'Martin Frobisher North-West Passage expedition 1576-78', rmg.co.uk/stories/topics/martin-frobisher-north-west-passage-expedition-1576-78.

T. Shabad, 'Soviet nuclear ship reaches North Pole through the arctic ice', *New York Times*, 18 August 1977, nytimes.com/1977/08/18/archives/soviet-nuclear-ship-reaches-north-pole-through-arctic-ice.html.

D. Simons, 'Arctic 30 jailed in Russia argue their case in European court', Greenpeace, 12 July 2018, greenpeace.org/international/story/18119/arctic-30-jailed-in-russia-argue-their-case-in-european-court/.

T. Tesi et al., 'Rapid Atlantification along the Fram Strait at the beginning of the 20th century', *Science Advances* 7, no. 48 (24 November 2021): doi.org/10.1126/sciadv.abj294.

A.N. Vylegzhanin, O.R. Young and P.A. Berkman, 'The Central Arctic Ocean Fisheries Agreement as an element in the evolving Arctic Ocean governance complex', *Marine Policy* 118, art. 104001 (2020): doi.org/10.1016/j.marpol.2020.104001.

CHAPTER 7: THE LAST FRONTIER

C.M. Brooks, 'Competing values on the Antarctic high seas: CCAMLR and the challenge of marine-protected areas', *Polar Journal* 3, no. 2 (2013): doi.org/10.1080/2154896X.2013.854597.

C.M. Brooks et al., 'Antarctic fisheries: factor climate change into their management', *Nature* 558 (14 June 2018): doi.org/10.1038/d41586-018-05372-x.

C.M. Brooks et al., 'The Ross Sea, Antarctica: A highly protected MPA in international waters', *Marine Policy* 134, art. 104795 (2021): doi.org/10.1016/j.marpol.2021.104795.

C.M. Brooks et al., 'Science-based management in decline in the Southern Ocean', *Science* 354, no. 6309 (14 October 2016): doi.org/10.1126/science.aah4119.

Deutsche Welle, 'Italy drops charges against Sea-Watch captain Carola Rackete charges', 23 December 2021, dw.com/en/italy-drops-charges-against-sea-watch-captain-carola-rackete/a-60245299.

M.J. Dunn et al., 'Population size and decadal trends of three penguin species nesting at Signy Island, South Orkney Islands', *PLOS ONE* 11, no. 10, art. e0164025 (2016): doi.org/10.1371/journal.pone.0164025.

J.T. Hinke et al., 'Individual variation in migratory movements of chinstrap penguins leads to widespread occupancy of ice-free winter habitats over the continental shelf and deep ocean basins of the Southern Ocean', *PLOS ONE* 14, no. 12, art. e0226207 (2019): doi.org/10.1371/journal.pone.0226207.

R. Huntford, *The Last Place on Earth: Scott & Amundsen's Race to the South Pole* (Abacus Books, 2000).

E. Kolbert, *The Sixth Extinction: An Unnatural History* (Henry Holt and Co., 2014).

L. Krüger et al., 'Antarctic krill fishery effects over penguin populations under adverse climate conditions: Implications for the management of fishing practices', *Ambio* 50 (2021): doi.org/10.1007/s13280-020-01386-w.

A. Lansing, *Endurance: Shackleton's Incredible Voyage* (Carroll & Graf Publishers, 1999).

N. Liu, C.M. Brooks and T. Qin (eds.), *Governing Marine Living Resources in the Polar Regions* (Edward Elgar Publishing, 2019).

S.R. Rintoul et al., 'Choosing the future of Antarctica', *Nature* 558 (2018): doi.org/10.1038/s41586-018-0173-4.

C. Roberts, *The Unnatural History of the Sea*, 6th ed. (Island Press, 2007).

Scientific Committee for the Conservation of Antarctic Marine Living Resources (CCAMLR), 'Domain 1 Marine Protected Area preliminary proposal part A-1: Priority areas for conservation', Report of the Thirty-Sixth Meeting of the Scientific Committee (16–20 October 2017), meetings.ccamlr.org/en/system/files?file=e-sc-xxxvi_0.pdf.

Scientific Committee for the Conservation of Antarctic Marine Living Resources (CCAMLR), 'Revised proposal for a conservation measure establishing a Marine Protected Area in Domain 1 (Western Antarctic Peninsula and South Scotia Arc)', Delegations of Argentina and Chile (October 2020), meetings.ccamlr.org/en/ccamlr-39/08-rev-1.

M. Siegert et al., 'The Antarctic Peninsula under a 1.5°C global warming scenario', *Frontiers in Environmental Science* 7 (28 June 2019): doi.org/10.3389/fenvs.2019.00102.

N. Strycker et al., 'A global population assessment of the chinstrap penguin (*Pygoscelis antarctica*)', *Scientific Reports* 10, art. 19474 (2020): doi.org/10.1038/s41598-020-76479-3.

R. Syal, 'Countries fail to agree on Antarctic conservation measures for fifth straight year', *Mongabay*, 15 November 2021, news.mongabay.com/2021/11/countries-fail-to-agree-on-antarctic-conservation-measures-for-fifth-straight-year/.

R. Syal, 'Licence to krill: The destructive demand for a "better" fish oil', *Guardian*, 7 September 2021, theguardian.com/environment/2021/sep/07/license-to-krill-the-destructive-demand-for-a-better-fish-oil.

Z.T. Sylvester and C.M. Brooks, 'Protecting Antarctica through co-production of actionable science: Lessons from the CCAMLR marine protected area process', *Marine Policy* 111, art. 103720 (2020): doi.org/10.1016/j.marpol.2019.103720.

W.Z. Trivelpiece et al., 'Variability in krill biomass links harvesting and climate warming to penguin population changes in Antarctica', *Proceedings of the National Academy of Sciences* 108, no. 18 (2011): doi.org/10.1073/pnas.1016560108.

G. Walker, *Antarctica: An Intimate Portrait of the World's Most Mysterious Continent* (Bloomsbury, 2012).

V. Warwick-Evans, 'Using habitat models for chinstrap penguins *Pygoscelis antarctica* to advise krill fisheries management during the penguin breeding season', *Diversity and Distributions* 24, no. 12 (2018): doi.org/10.1111/ddi.12817.

CHAPTER 8: GENES, DRUGS AND JUSTICE

W. Appeltans et al., 'The magnitude of global marine species diversity', *Current Biology* 22, no. 23 (2012): doi.org/10.1016/j.cub.2012.09.036.

S. Arnaud-Haond, J.M. Arrieta and C.M. Duarte, 'Marine biodiversity and gene patents', *Science* 331, no. 6024 (2011): doi.org/10.1126/science.1200783.

R. Blasiak et al., 'Corporate control and global governance of marine genetic resources', *Science Advances* 4, no. 6 (6 June 2018): doi.org/10.1126/sciadv.aar52.

R. Blasiak et al., 'The ocean genome and future prospects for conservation and equity', *Nature Sustainability* 3 (2020): doi.org/10.1038/s41893-020-0522-9.

S.R. Bown, *Scurvy: How a Surgeon, a Mariner, and a Gentleman Solved the Greatest Medical Mystery of the Age of Sail* (St. Martin's Press, 2005).

A. Broggiato et al., 'Mare Geneticum: Balancing governance of marine genetic resources in international waters', *International Journal of Marine and Coastal Law* 33, no. 1 (2018): doi.org/10.1163/15718085-13310030.

M. Elmer and L. Fimbres Wood, 'Deep sea anti-cancer drug discovered by Scripps scientists enters final phase of clinical trials', Scripps Institution of Oceanography, UC San Diego, 7 April 2020, scripps.ucsd.edu/news/deep-sea-anti-cancer-drug-discovered-scripps-scientists-enters-final-phase-clinical-trials.

W. Fenical, 'Marine microbial natural products: The evolution of a new field of science', *Journal of Antibiotics* 73 (2020): doi.org/10.1038/s41429-020-0331-4.

K. Gould, 'Antibiotics: From prehistory to the present day', *Journal of Antimicrobial Chemotherapy* 71, no. 3 (2016): doi.org/10.1093/jac/dkv484.

O. Heffernan, 'Why a landmark treaty to stop ocean biopiracy could stymie research', *Nature* 580, no. 7801 (2020): doi.org/10.1038/d41586-020-00912-w.

L. Nordling, 'Rooibos tea profits will be shared with Indigenous communities in landmark agreement', *Nature* 575, no. 7781 (2019): doi.org/10.1038/d41586-019-03374-x.

G. Posner, *Pharma: Greed, Lies, and the Poisoning of America* (Avid Reader Press, 2021).

M. Rabone et al., 'Access to marine genetic resources (MGR): Raising awareness of best-practice through a new agreement for biodiversity beyond national jurisdiction (BBNJ)', *Frontiers in Marine Science* 6 (12 September 2019): doi.org/10.3389/fmars.2019.00520.

V. Shiva, *Biopiracy: The Plunder of Nature and Knowledge* (South End Press, 1999).

C.A. Suttle, 'Marine viruses – major players in the global ecosystem', *Nature Reviews Microbiology* 5 (2007): doi.org/10.1038/nrmicro1750.

M. Vierros et al., 'Who owns the ocean? Policy issues surrounding marine genetic resources', *Bulletin Limnology and Oceanography* 25, no. 2 (May 2016): doi.org/10.1002/lob.10108.

S. Yooseph et al., 'The *Sorcerer II* Global Ocean Sampling Expedition: Expanding the universe of protein families', *PLOS Biology* 5, no. 3, art. e16 (2007): doi.org/10.1371/journal.pbio.0050016.

CHAPTER 9: DEEP TROUBLE

V. De Lucia, 'Splashing down the International Space Station in the Pacific Ocean: Safe disposal or trashing the ocean commons?' *EJIL:Talk!*, blog of the *European Journal of International Law*, 23 February 2022, ejiltalk.org/splashing-down-the-international-space-station-in-the-pacific-ocean-safe-disposal-or-trashing-the-ocean-commons/.

V. De Lucia and V. Iavicoli, 'From outer space to ocean depths: The spacecraft cemetery and the protection of the marine environment in areas beyond national jurisdiction', *California Western International Law Journal* 49, no. 2 (2019): doi.org/10.2139/ssrn.3153458.

T. Evanisko, 'Bilge dumping: What it is, why you should care, and what can be done', SkyTruth Report, May 2020, skytruth.org/wp-content/uploads/2020/10/SkyTruth-Global-Bilge-Dumping-Report.pdf.

GESAMP Joint Group of Experts on the Scientific Aspects of Marine Environmental Protection, 'Pollution in the open ocean: A review of assessments and related studies', *GESAMP Reports and Studies* 79 (August 2009): gesamp.org/site/assets/files/1261/pollution-in-the-open-oceans-a-review-of-assessments-and-related-studies-en.pdf.

K. Gilardi (ed.), 'Sea-based sources of marine litter', GESAMP Joint Group of Experts on the Scientific Aspects of Marine Environmental Protection Working Group 43, *GESAMP Reports and Studies* 108 (2021): gesamp.org/site/assets/files/2213/rs108e.pdf.

Guardian, 'Mir today: Safe splashdown for space station', 23 March 2001, theguardian.com/science/2001/mar/23/spaceexploration.

L.E. Haram et al., 'Emergence of a neopelagic community through the establishment of coastal species on the high seas', *Nature Communications* 12, art. 6885 (2021): doi.org/10.1038/s41467-021-27188-6.

D.M. Harland, *The Story of Space Station Mir* (Springer, 2005).

R. Khatchadourian, 'The elusive peril of space junk', *New Yorker*, 28 September 2020, newyorker.com/magazine/2020/09/28/the-elusive-peril-of-space-junk.

L. Lebreton et al., 'Industrialised fishing nations largely contribute to floating plastic pollution in the North Pacific subtropical gyre', *Scientific Reports* 12, art. 12666 (2022): doi.org/10.1038/s41598-022-16529-0.

L. Mead, '"The ocean is not a dumping ground": Fifty years of regulating ocean dumping', International Institute for Sustainable Development, Brief 28 (December 2021): iisd.org/system/files/2021-11/still-one-earth-ocean-dumping.pdf.

R. Murtazin et al., 'ISS controlled deorbit: Challenges and solutions', paper presented at the 68th International Astronautical Congress (IAC), Adelaide, Australia (25–29 September 2017): researchgate.net/publication/340902025.

B.C. O'Leary et al., 'Options for managing human threats to high seas biodiversity', *Ocean & Coastal Management* 187, art. 105110 (2020): doi.org/10.1016/j.ocecoaman.2020.105110.

P.E. Tyler, 'Mir space station sizzles to ending over Pacific', *New York Times*, 23 March 2001, nytimes.com/2001/03/23/world/mir-space-station-sizzles-to-ending-over-pacific.html.

P.E. Tyler, 'Russians find pride, and regret, in Mir's splashdown', *New York Times*, 24 March 2001, nytimes.com/2001/03/24/world/russians-find-pride-and-regret-in-mir-s-splashdown.html.

S. Zhang et al., 'Challenges and countermeasures for international ship waste management: IMO, China, United States, and EU', *Ocean & Coastal Management* 213, no. 3, art. 105836 (2021): doi.org/10.1016/j.ocecoaman.2021.105836.

CHAPTER 10: THE COLD RUSH

A. Alshehi, *Filling the Empty Quarter: Declaring a Green Jihad on the Desert* (self-pub., 2015).

J. Amos, 'A68: "Megaberg" dumped huge volume of fresh water', BBC News, 20 January 2022, bbc.com/news/science-environment-60060299.

A. Braakmann-Folgmann et al., 'Observing the disintegration of the A68A iceberg from space', *Remote Sensing of the Environment* 270, art. 112855 (1 March 2022): doi.org/10.1016/j.rse.2021.112855.

S.E. Bruneau and K. Redmond, *Iceberg Alley: A Journal of Nature's Most Awesome Migration* (Flanker Press, 2010).

A. Condron, 'Towing icebergs to arid regions to reduce water scarcity', *Scientific Reports* 13, art. 365 (2023): doi.org/10.1038/s41598-022-26952-y.

A. Costov and J. Applemann, 'Exploitation of Antarctic iced freshwater: A call to unfreeze legal discourse', *Groningen Journal of International Law: Open Issue* 9, no. 1 (2021): doi.org/10.21827/GroJIL.9.1.60-77.

Euronews, 'Will an Emirati businessman succeed in towing an iceberg to the UAE?' 5 July 2019, euronews.com/2019/07/05/will-an-emirati-businessman-succeed-in-towing-an-iceberg-to-the-uae.

O. Heffernan, 'Towing icebergs to Cape Town is a poor way to halt water crisis', *New Scientist*, 8 May 2018, newscientist.com/article/2168339-towing-icebergs-to-cape-town-is-a-poor-way-to-halt-water-crisis/.

C. Holden, 'Experts ponder icebergs as relief for world water dilemma', *Science* 198, no. 4314 (October 1977): doi.org/10.1126/science.198.4314.274.

Z. Karimidastenaei et al., 'Polar ice as an unconventional water resource: Opportunities and challenges', *Water* 13, no. 22, art. 3220 (2021): doi.org/10.3390/w13223220.

S. Kirchner et al., 'Harvesting icebergs and space rocks: Small scale resource extraction entrepreneurship in areas beyond national jurisdictions', Arctic Governance Research Group, Arctic Centre, University of Lapland, Rovaniemi, Finland (2020), er.nau.edu.ua/handle/NAU/48771.

A. Lam and M. Tegelberg, 'Dark tourism in Iceberg Alley: The hidden ecological costs of consuming iceberg deaths', in *Criminal Anthropocenes: Media and Crime in the Vanishing Arctic* (Palgrave Macmillan, 2020), doi.org/10.1007/978-3-030-46004-4_5.

N. Malan, 'Are icebergs a realistic option for augmenting Cape Town's water supply?' *Water Wheel* 17, no. 2 (March/April 2018): wrc.org.za/wp-content/uploads/mdocs/ww_March_April2018_web.pdf.

T. Smedley, 'The outrageous plan to haul icebergs to Africa', BBC Future, 21 September 2018, bbc.com/future/article/20180918-the-outrageous-plan-to-haul-icebergs-to-africa.

CHAPTER 11: PARADISE LOST

D. Abulafia, *The Great Sea: A Human History of the Mediterranean* (Penguin, 2014).

S.B. Borrelle et al., 'Why we need an international agreement on marine plastic pollution', *Proceedings of the National Academy of Sciences* 114, no. 38 (19 September 2017): doi.org/10.1073/pnas.1714450114.

E. Cassidy and A. Snyder, 'Ships spend hundreds of thousands of hours a year trawling Europe's Marine Protected Areas', *Resource Watch*, 28 February 2019, blog.resourcewatch.org/2019/02/28/ships-spend-hundreds-of-thousands-of-hours-a-year-trawling-in-europes-marine-protected-areas/.

E. D'Angelo, 'Sweep island, an artificial island for the environment', *Collater.al*, June 2021, collater.al/en/sweep-island-artificial-island-for-envirinment-architecture/.

L. David, N. Di-Meglio and P. Monestiez, 'Sperm whale ship strikes in the Pelagos Sanctuary and adjacent waters: Assessing and mapping collision risks in summer', IWC *Journal of Cetacean Research and Management* 18, no. 18 (2019): doi.org/10.47536/jcrm.v18i1.446.

C.M. Duarte et al., 'The soundscape of the Anthropocene ocean', *Science* 371, no. 6529 (5 February 2021): doi.org/10.1126/science.aba4658.

M. Dureuil et al., 'Elevated trawling inside protected areas undermines conservation outcomes in a global fishing hot spot', *Science* 362, no. 6421 (21 December 2018): doi.org/10.1126/science.aau0561.

R. Ellis, *The Great Sperm Whale: A Natural History of the Ocean's Most Magnificent and Mysterious Creature* (University Press of Kansas, 2011).

M.C. Fossi et al., 'The Pelagos Sanctuary for Mediterranean marine mammals: Marine Protected Area (MPA) or marine polluted area? The case study of the striped dolphin (*Stenella coeruleoalba*)', *Marine Pollution Bulletin* 70, nos. 1–2 (March 2013): doi.org/10.1016/j.marpolbul.2013.02.013.

M.C. Fossi et al., 'Plastic debris occurrence, convergence areas and fin whales feeding ground in the Mediterranean Marine Protected Area Pelagos Sanctuary: A modeling approach', *Frontiers in Marine Science* 4 (2017): doi.org/10.3389/fmars.2017.00167.

F. Grossi et al., 'Locating ship strike risk hotspots for fin whale (*Balaenoptera physalus*) and sperm whale (*Physeter macrocephalus*) along main shipping lanes in the North-Western Mediterranean Sea', *Ocean & Coastal Management* 212, art. 105820 (2021): doi.org/10.1016/j.ocecoaman.2021.105820.

E. Kleverlaan et al., 'Strategic guidance document on how to prepare a successful PSSA proposal to IMO', WWF Mediterranean Marine Initiative (2021), wwfeu.awsassets.panda.org/downloads/wwf_pssa_guidance_document_2021_final.pdf.

A. Maglio et al., 'Overview of the noise hotspots in the ACCOBAMS area Part I – Mediterranean Sea', paper presented at the Sixth Meeting of the Parties to ACCOBAMS, Monaco (22–25 November 2016), accobams.org/wp-content/uploads/2020/01/MOP6.Doc28Rev1_Overview_noise_hot_spots_-ACCOBAMS_area_Part_Mediterranean.pdf.

G. Notarbartolo-di-Sciara et al., 'The Pelagos Sanctuary for Mediterranean marine mammals', *Aquatic Conservation: Marine and Freshwater Ecosystems* 18, no. 4 (June 2008): doi.org/10.1002/aqc.855.

L. Nunny (ed.), 'Under pressure: The need to protect whales and dolphins in European waters', OceanCare report (April 2021), oceancare.org/wp-content/uploads/2022/11/Animal_Species_Protection_Under-Pressure_Whales-and-Dolphins_EU_Report_OceanCare_EN_146p_2021.pdf.

S. Panigada et al., 'Mediterranean fin whales at risk from fatal ship strikes', *Marine Pollution Bulletin* 52, no. 10 (October 2006): doi.org/10.1016/j.marpolbul.2006.03.014.

L. Rendell and A. Frantzis, 'Mediterranean sperm whales, *Physeter macrocephalus*: The precarious state of a lost tribe', in *Advances in Marine Biology: Mediterranean Marine Mammal Ecology and Conservation*, vol. 75, ed. G. Notarbartolo Di Sciara, M. Podestà and B.E. Curry (Academic Press, 2016), doi.org/10.1016/bs.amb.2016.08.001.

M. Roscam Abbing, *Plastic Soup: An Atlas of Ocean Pollution* (Island Press, 2019).

J. Steinbeck and E.F. Ricketts, *Sea of Cortez: A Leisurely Journal of Travel and Research* (Paul P. Appel, Publisher, 1971).

M. Streit-Bianchi, M. Cimadevila and W. Trettnak (eds.), *Mare Plasticum – The Plastic Sea: Combatting Plastic Pollution Through Science and Art* (Springer, 2020).

30×30: A Blueprint for Ocean Protection (University of Oxford, University of York and Greenpeace, 2019): greenpeace.org/30x30.

T. Yi et al., 'International cooperation and coordination in the global legislation of high seas ABMTs including MPAs: Taking OSPAR practice as reference', *Marine Policy* 133, no. 5, art. 104767 (2021): doi.org/10.1016/j.marpol.2021.104767.

CHAPTER 12: HOPE FOR THE HIGH SEAS

Deep Sea Conservation Coalition: savethehighseas.org

High Level Panel for a Sustainable Ocean Economy: oceanpanel.org

High Seas Alliance: highseasalliance.org

Intergovernmental Conference on an international legally binding instrument under the United Nations Convention on the Law of the Sea on the conservation and sustainable use of marine biological diversity of areas beyond national jurisdiction: un.org/bbnj

Port State Measures Agreement: fao.org/port-state-measures/en

Sylvia Earle's Mission Blue: mission-blue.org/

UN Sustainable Development Goals: sdgs.un.org/goals

United Nations Framework Convention on Climate Change: unfccc.int/

World Trade Organization Agreement on Fisheries Subsidies: wto.org/english/tratop_e/envir_e/session_14_agreement_on_fisheries_subsidies.pdf

FURTHER READING

Clover, C. *The End of the Line: How Overfishing Is Changing the World and What We Eat*. New Press, 2006.

Corfield, R. *The Silent Landscape: The Scientific Voyage of* HMS *Challenger*. Joseph Henry Press, 2003.

Dodds, K., and M. Nuttall. *The Scramble for the Poles: The Geopolitics of the Arctic and Antarctic*. Polity, 2015.

Kunzig, R. *Mapping the Deep: The Extraordinary Story of Ocean Science*. W.W. Norton, 2000.

Pauly, D. *Vanishing Fish: Shifting Baselines and the Future of Global Fisheries*. Greystone Books, 2019.

Roberts, C. *The Ocean of Life: The Fate of Man and the Sea*. Penguin Books, 2013.

Roberts, C. *The Unnatural History of the Sea*, 6th ed. Island Press, 2007.

Rogers, A. *The Deep: The Hidden Wonders of Our Oceans and How We Can Protect Them*. Wildfire, 2020.

Rozwadowski, H.M. *Fathoming the Ocean: The Discovery and Exploration of the Deep Sea*. Belknap, 2008.

Urbina, I. *The Outlaw Ocean: Journeys Across the Last Untamed Frontier*. Knopf, 2019.

INDEX

Abdulaziz bin Salman Al Saud, 309
acidity, ocean, 130–31
actinomycetes, 206–7, 208
Aegean Sea, 11
Aequorea victoria (jellyfish), 217
Agnes na Mara (bottom trawler), 7–8
Agreement on the Conservation of Cetaceans in the Black Sea, Mediterranean Sea and contiguous Atlantic area (ACCOBAMS), 296
Agulhas Current, 250
AIS (Automated Information System) beacons, 75–76, 77, 85, 87
Alaska pollock, 83
Albert I (prince of Monaco), 273
algae, sea-ice, 175–76
Almås, Karl Andreas, 66
Alshehi, Abdulla, 254, 255, 256, 260–61, 262
Amon, Diva, 95–96, 96–97, 115
Amos, John, 75, 76
Amundsen, Roald, 166
anchoveta (Peruvian anchovy), 40, 81–82
Antarctica, 190, 194–95. *See also* Southern Ocean
Antarctic Treaty, 196
antibiotics, 207, 209, 227. *See also* medicines
Apollo II (deep-sea mining machine), 103, 104, 105, 106–7, 111–12, 113–14, 125
Appelmann, Jessica, 263
aquaculture: fish feed, 44–46, 48–49; salmon, 38–39, 46–50
Arabian Sea, 58–59
Arctic: Arctic 30 (Greenpeace protesters), 168–70; Atlantification of, 167; attempts to cross, 165–66; Chinese shipping plans in Central Passage, 25, 158–60; climate change in, 161–62, 182; Fram Strait, 154, 155, 159–60, 162, 166, 171–72, 175; Greenpeace expedition, 161, 162–65, 167–68, 172–74, 175–76; High Seas Treaty and, 304; as ice-dependent ecosystem, 164–65, 175; Lomonosov Ridge, 177–78, 180; lucrative resources under, 178; Russian claims and plans, 170, 178–79; seabed claims and governance, 178–82; sea ice, 25, 154–56, 161, 162, 167, 171–72, 174; shipping impacts, 160–61
Arctic Ice Project, 153–56
Arctic Sunrise (Greenpeace ship), 164, 167–69, 172–73, 175–76, 188, 189–90, 199–201
Argentina, 202
Arktika (Soviet icebreaker), 166
Atlantic Ocean, 56, 167
Atolla (*Atolla wyvillei*), 67
Automated Information System (AIS) beacons, 75–76, 77, 85, 87
azoic theory, 11–12
AZT, 28, 210

Bainimarama, Frank, 124
Bandit Six, 71–72
Barcelona Convention, 296
Barents Sea, 20, 41, 249
BarentsWatch, 49

Barron, Gerard, 118–19
Barton, Otis, 52
Basel Convention, 236
BASF, 222, 223
Bathysphere (submersible), 52–53
Beagle Channel, 188
BECCS (bioenergy with carbon capture and storage), 134
Beebe, William, 52–53
beetle, rove, 214
Belgium, 235
Benguela Current, 250
Bergman, Bjorn, 69, 73–74, 75–79, 79–80, 84–86, 89, 307
Bergmann, Werner, 210, 211
Berg Water, 264–65
Beyond Plastic Med (BeMed) initiative, 294
bigeye, 39
bilge water, 240–41
billfish, 63, 106
biodiscovery: actinomycetes, 206–7; biopiracy, 218–19, 220–21, 223–24, 226, 227; concerns and dilemmas, 224; early pioneers, 211–14; marine biotechnology industry, 28, 211; marine genetic resources, 215–17, 219–21, 222–23, 225–26, 303–4; medicines, 27–28, 205, 207, 208, 209–11, 217, 221–22, 227; microbes, 205–7, 208–9, 214–16; regulations, 223–24, 226
bioenergy with carbon capture and storage (BECCS), 134
bio-inspired products, 226
biological carbon pump, 51
BioMar, 44–46
biopiracy, 218–19, 220–21, 223–24, 226, 227. *See also* biodiscovery
bioprospecting, 210, 218, 227. *See also* biodiscovery
Birkeland AS, 35
Birkhold, Matthew H., 256, 264
black scabbard, 7
Blasiak, Robert, 217, 222, 224–25
Blue Acceleration, 28–29, 310
blue whiting, 35–36, 36–37, 43–44, 45–46, 62
boundary currents, 250
BP, 75
bristlemouth, 13, 33
Britain. *See* United Kingdom
brittle stars, 95, 281
Bruneau, Stephen, 256, 268–70
Brunel, Olivier, 165
bubble raft snail (violet sea snail), 251
Buesseler, Ken, 34, 50–51, 53–54, 68
Burreson, Jay, 213

Cabo Pulmo, 281–82
Calanus, 40
Caldeira, Ken, 146
Cameron, James, 34
Canada, 25, 125, 159, 177, 178, 180, 181, 221
cannon-shot rule, 20
Cape, Mattias, 167
Cape Horn, 188
Cape Town, 261–62
Captura, 143
carbon storage: and artificial upwelling and downwelling, 141–42; BECCS (bioenergy with carbon capture and storage), 134; biological carbon pump, 51; carbon dioxide removal (CDR) technologies, 133–34, 151; ecosystem restoration and, 140–41; electrogeochemistry and, 142–43; enhanced weathering and, 135–36, 137–38; impacts on ocean, 27, 130–31; marine snow and, 51, 53–54, 61; ocean fertilisation and, 126–29, 131–32, 139, 144–45; ocean's capacity for, 50, 129–30; in Saya de Malha Bank, 3; seaweed cultivation and, 139–40; solubility pump and, 54; in twilight zone, 24, 33–34, 39, 50–51, 53–54. *See also* climate interventions
Carta Marina (map), 4
cephalopods, 82–83. *See also* octopus; squid
Chakravarty, Sid, 77
Challenger Deep, 13–14
Challenger expedition, 12–14, 52, 97–98
Chapsos, Ioannis, 73
Chile, 90, 184, 202, 259
Chilean Antarctic Institute, 198
China: Arctic fishing agreement, 181–82; Arctic shipping plans, 25, 158–60;

distant-water (squid) fishing fleet, 74, 79–80, 83–84, 84–85, 89, 90–91; krill fishery, 184, 185; plastics from, 244; Senkaku Islands dispute, 108; Southern Ocean sanctuary and, 202
chinstrap penguins, 185–86, 191–94, 196, 198, 199, 200–201
chlorophyll, 57
Chopra, Govinder Singh, 120
circular economy, 113, 118
Clarion-Clipperton Zone (CCZ), 92, 95, 96–97, 105, 109–10, 117, 124
Clean Water Act (US), 236
climate change: in Arctic, 161–62, 182; author's work on, 306–7; dead zones and, 131, 242; Framework Convention on Climate Change, 305; greenhouse gas emissions, 30, 130, 132–33; High Seas Treaty and, 304; individual responsibility for, 311; minerals and metals needed for renewable energy, 101–2; ocean as buffer, 30; Oxygen Minimum Zones and, 58; Paris Climate Accord, 132, 305, 311, 313; sea ice and, 25–26, 154–56, 161, 162, 167, 171–72, 174; Southern Ocean and, 185, 198. *See also* climate interventions
Climate Foundation, 143
climate interventions: artificial upwelling and downwelling, 141–42; assessment of options by NASEM, 138–42; in atmosphere, 151; BECCS (bioenergy with carbon capture and storage), 134; carbon dioxide removal (CDR) technologies, 133–34, 151; challenges facing, 131, 144–45; ecosystem restoration, 140–41; electrogeochemistry, 142–43; enhanced weathering, 135–36, 137–38; funding and support for, 144, 156–57; marine cloud brightening (solar albedo modification), 146–47, 148, 149–51, 152–53; Musk's XPRIZE for, 135, 143; necessity and interest in, 134–35, 145; ocean fertilisation, 126–29, 131–32, 139, 144–45; sea-ice restoration, 153–56; seaweed cultivation, 139–40; Wanser on, 145–47, 150–51, 152, 156, 157
ClimateWorks, 138
Climitigation, 136–37
clouds, 146–50, 152–53
cloud seeding, 151, 152
Coalition Against Biopiracy, 218–19
Coanda effect, 107
cobalt, 101, 102, 104, 118
cobalt-rich crusts, 94, 95
cod, 20, 40, 311; Arctic, 211, 216
Cod Wars, 20–21
Cold War, 151, 184, 231
Coleridge, Samuel Taylor: *The Rime of the Ancient Mariner*, 265
Collison, Patrick and John, 144
Commission for the Conservation of Antarctic Marine Living Resources (CCAMLR), 184, 196–97, 198, 199, 201, 202, 303
Commission on the Limits of the Continental Shelf (CLCS), 179
common dolphinfish (mahi-mahi), 81, 82
commons, tragedy of the, 26–27
Condron, Alan, 261–62
Congo, 104, 109
conservation, 312–13. *See also* High Seas Treaty; marine protected areas
continental shelf, 21–22, 180
continental slope, 59
Convention on Biological Diversity (CBD), 129, 274
Cook, James, 166
Cook Islands, 122
copper, 95, 101, 113, 118, 178
Costa Rica Thermal Dome, 57–58, 304
Costov, Ana, 263
crabs, 50, 199
Crespo, Guillermo Ortuño, 63–64
criminality. *See* illegal, unreported and unregulated (IUU) fishing
CSIC (Spanish National Research Council), 104
currents: flotsam circulated by, 245–48, 249, 251; gyres, boundary currents and eddies, 250–51; thermohaline circulation (global conveyor belt), 248–49
Currie, Duncan, 123, 301
cuttlefish, 82

Dalkey Sound, 314
Darwin Centre (Natural History Museum, London), 95
Dassault Systèmes, 260
Davis, Keith, 75
DDT, 236, 237
dead zones, 26–27, 131, 242
deep sea, 11–14, 96. *See also* Clarion-Clipperton Zone; high seas; mining, deep-sea
Deep Sea Conservation Coalition, 301, 302
Deepwater Horizon oil spill, 75, 238
de Jonge, Laurens, 104–5, 112
De Lucia, Vito, 233–34, 235
Denmark, 177, 178, 180, 181
Descartes, René, 148
de Stigter, Henko, 102–3, 104, 105–6, 107, 108–9, 112, 113, 124, 125, 307
diatoms, 55–56
digital sequence information (DSI), 217, 222, 226
DISCOL (disturbance and recolonisation), 111
Dodds, Klaus, 179, 180–81
dolphins, 58, 77, 128, 207, 273, 313; common, 284; Risso's, 188, 294; striped, 290, 297
dragonfish, 33
Drake Passage, 188, 189
drift nets, 77, 280, 294, 312–13
Duarte, Carlos, 65
dumping. *See* pollution and ocean dumping
Duncan, Kate, 205, 206–7, 208, 211, 227
Dutch East India Company, 10, 19

Earle, Sylvia, 274
Ebbesmeyer, Curtis, 245–46
Ecologically and Biologically Significant Areas (EBSAS), 274
ecosystems restoration, 140–41
Ecuador, 78–79, 84, 88, 129, 223
eddies, 250–51
eels, 2, 59
electrogeochemistry, 142–43
Elephant Island, 183, 186, 190–91, 192–96, 197, 198, 201, 202, 204
El Niño, 82
Elvestad, Roger, 45
Elysia rufescens, 213
Emerson, Ralph Waldo, 147–48
enhanced weathering, 135–36, 137–38
environmental impact assessments (EIAS), 302–3, 308
Environmental Justice Foundation, 74
Environmental Protection Agency, 129
Esperanza (Greenpeace ship), 162–63, 164, 167–68, 173, 182, 187–88, 189–90
euphotic zone, 56
European Maritime, Fisheries and Aquaculture Fund (EMFAF), 309
European Union (EU), 21, 36, 43, 83, 181, 196, 220, 301, 309
Exclusive Economic Zones (EEZS), 21, 179–80, 290, 311
ExxonMobil, 169, 244
Exxon Valdez oil spill, 238

Faroe Islands, 36
Fenical, Bill, 207–8, 227
Field, Leslie Ann, 153–56
Fiji, 124
Finland, 165, 178
fin whales, 128, 184, 273, 279, 284, 290–91, 293, 294, 313
FishCRIME conference, 69–70
fisheries: *Agnes na Mara*'s offshore experience, 7–8; Arctic ban, 181–82; bottom trawling, 8–9, 20–21, 29, 64, 93, 279–80, 300–301, 302; drift nets, 77, 280, 294, 312–13; European Union and, 309; High Seas Treaty and, 304; history of high-seas fishing, 40–41; krill, 25–26, 184–85, 196–99, 202; longlines, 63, 64, 76; management issues, 23, 62–63, 75; mesopelagic (twilight zone), 24, 34–35, 42–43, 61–62, 65–68; Norwegian mesopelagic fishery, 24, 34–35, 35–36, 39–40, 41, 43–44, 68; Norwegian overfishing disputes, 36–37, 45; observers, 74–75; overfishing, 63–64, 84; positioning,

43–44; proposal to close high seas to, 311–12; regional fisheries management organisations (RFMOs), 23, 62–63, 304; salmon farming, 38–39, 46–50; squid, 26, 64, 74, 80–81, 82–83, 87–88, 90–91; tragedy of the commons and, 26; as unprofitable and unnecessary, 64–65. *See also* illegal, unreported and unregulated (IUU) fishing
fish feed, 44–46, 48–49
flags of convenience (open registers), 86–87, 160, 308
Fleming, Alexander, 209
Flores Man (*Homo floresiensis*, 'the hobbit'), 9
flotsametrics, 245–47, 249, 251
flying fish, 3
Food and Agriculture Organization (FAO), 83, 185
Forbes, Edward, 11
forkbeard, 7
Forrest, Steve, 193, 194, 203
Forster, Johann, 3–4
Framework Convention on Climate Change, 305
Fram Strait, 154, 155, 159–60, 162, 166, 171–72, 175
France, 20, 117, 125, 235, 296
Franklin, John, 166
free seas, 16–17, 19–20, 98, 220
fried egg sea slug, 212–13
Frobisher, Martin, 165–66
Fu Yuan Yu Leng 999 (Chinese reefer), 78–79, 90

Gagne, Wallace, 269
Gakkel Ridge, 2, 304, 306
Galápagos Islands, 78–79, 90, 129
GenBank, 217, 219, 222
General Fisheries Commission for the Mediterranean (GFCM), 296
genetic resources: concerns and dilemmas, 224; corporate control over, 221–23; gene sequencing, 214–17; High Seas Treaty and, 303–4; from the ocean, 215–17, 219–21, 222–23, 225–26, 303–4
geoengineering. *See* climate interventions
George, Russ, 126–29, 131–32, 139, 152
Germany, 20, 235
Gianni, Matthew, 301
giant red mysid, 59
Gilead Sciences, 221–22
Gjerde, Kristina, 300, 301
Glandon, Hillary, 163, 165, 175–76
global conveyor belt (thermohaline circulation), 248–49
Global Fishing Watch (GFW), 75–79, 79–80, 85–86, 87–90, 311. *See also* SkyTruth
Global Sea Mineral Resources (GSR), 114
Global Seed Vault, 161
Glover, Adrian, 96–97
gold, 95, 113, 178
Great Acceleration, 28
Great Age of Discovery, 10
Great Barrier Reef, 153, 280
Great Pacific Garbage Patch, 243, 244, 250, 252, 305
greenhouse gas emissions, 30, 130, 132–33. *See also* carbon storage; climate change; climate interventions
Greenland, 167, 255, 257, 263
Greenland halibut, 7
Greenland Sea, 249
Greenpeace: Arctic 30 protesters, 168–70; Arctic expedition, 161, 162–65, 167–68, 172–74, 175–76; individual responsibility to support, 311; ocean fertilisation and, 129; Southern Ocean expedition, 186, 187–88, 189–91, 192–96, 199–201, 203–4
Greenwich Island, 203
grenadier, 7
Grimaldo, Eduardo, 37, 40, 43, 44, 65–66, 67
Grøntvedt, Ove and Sivert, 46, 48
Grossi, Rafael Mariano, 253
Grotius, Hugo, 16–19; *On the Law of War and Peace*, 18; *Mare Liberum*, 16–17, 19–20
Gulf Stream, 131, 250
gummy squirrel, 95, 108
Gundersen, Vidar, 44–45, 46
gyres, 250

hadal zone, 60
Haida Gwaii, 126–29, 131–32
hake, 81
Halaven, 28, 210–11
Half-Earth, 312
halibut, Greenland, 7
Hansa Carrier (freighter), 246
hatchetfish, 52
herring, 20, 35, 36–37, 39, 45
High Level Panel for a Sustainable Ocean Economy, 313
high seas: about, 1–2, 5, 22–23, 29–31; activism and protections, 30–31, 301, 302–6; author's fascination with the ocean, 5–7, 8; author's hopes for, 306–8, 313–14; benefits from offshore expansion, 27–28; Blue Acceleration, 28–29, 310; conservation successes, 312–13; exploration of, 4, 11–14, 221; free seas concept, 16–17, 19–20, 98, 220; governance issues, 23–24, 308–10; history of offshore adventures, 9–11; iconic places within, 2–3; individual responsibility for, 310–11; in popular imagination, 3–4; tragedy of the commons and, 26–27; United Nations Convention on the Law of the Sea (UNCLOS), 21, 22, 23, 219–20, 226. *See also* Arctic; biodiscovery; climate interventions; fisheries; icebergs; illegal, unreported and unregulated (IUU) fishing; marine protected areas; mining, deep-sea; ocean; pollution and ocean dumping; Southern Ocean
High Seas Alliance, 301, 311
High Seas Treaty, 302–4, 305–6, 309, 310, 311, 312, 313
Hitra Fish Museum, 47
Homo floresiensis (Flores Man, 'the hobbit'), 9
Hong Kong, 244
Hope Spots, 274
Howes, Rupert, 37
human rights abuses, 72–73, 74, 85, 308–9
Humboldt Current, 250
humpback whales, 184, 198, 203, 284, 313
hydrothermal vents, 2, 94–95, 119, 211, 223, 225

ice. *See* sea ice
icebergs: A-68 iceberg, 266, 270; harvesting for drinking water, 254–59, 260–64; IcebergFinder, 265; luxury water brands from, 264–65; marine impacts from melting, 266–67; in popular imagination, 265–66; towing feasibility, 267–71; towing history, 259–60
Iceland, 20–21, 36, 42–43, 178
IFCB (Imaging FlowCytobot), 55
illegal, unreported and unregulated (IUU) fishing (pirate fishing): about, 24–25, 69–70; Bandit Six, 71–72; Chinese distant-water fleets, 74, 79–80, 83–84, 84–85, 89, 90–91; crime convergence, 25, 70–71; flags of convenience and, 86–87, 160, 308; global efforts against, 90–91; Global Fishing Watch against, 75–79, 79–80, 85–86, 87–90, 311; going dark, 71, 76; High Seas Treaty and, 304; human rights abuses, 72–73, 74, 85, 308–9; Port State Measures Agreement and, 305; tracking methods, 75–76, 85–86, 87–88; transhipment, 85, 90
Imaging FlowCytobot (IFCB), 55
Important Bird and Biodiversity Areas (IBAS), 274
Important Marine Mammal Areas (IMMAS), 274
India, 42, 117, 218
Indian Ocean, 3, 26, 58, 62, 250
Ingraham, Jim, 246
In-Situ Ichthyoplankton Imaging System (ISIIS), 55
Institute for Sustainable Futures, 102
Inter-American Tropical Tuna Commission, 62
Intergovernmental Maritime Consultative Organization, 239. *See also* International Maritime Organization
Intergovernmental Panel on Climate Change (IPCC), 132, 133–34

International Commission for the Conservation of Atlantic Tunas (ICCAT), 296
International Convention for the Prevention of Pollution from Ships (MARPOL), 236
International Maritime Organization (IMO), 129, 160, 236, 239, 296, 304
International Seabed Authority (ISA), 23, 94, 102, 115–18, 122, 304, 308
International Space Station (ISS), 229, 232
International Union for Conservation of Nature (IUCN), 129, 293–94, 300
Iran, 42
Ireland, 7–8, 37, 314
Irigoien, Xabier, 65
Isaacs, John, 259–60
ISIIS (In-Situ Ichthyoplankton Imaging System), 55
Italy, 235, 296

Japan, 108, 184, 220, 221, 235, 244, 251, 253
Japanese pilchard, 83
Jaspars, Marcel, 206
jellyfish, 52, 59, 60, 67, 95, 106, 217
Jensen, Paul, 207–8, 209, 215–16, 227
Johannes, Bob, 212
Johnson, Samuel, 190

Kahalalide F, 213, 217
Kalas, Peggy, 301
Kean, Ed, 257–59, 263, 264, 265
Kellner, Rupert, 214
KhoiSan peoples, 218, 219
King, David, 141
Kiribati, 117, 121, 122
Knudsen, Christian, 176–77
krill, 25–26, 38, 184–85, 192, 196–99, 202
Kroodsma, David, 88
Kuroshio Current, 250

Lam, Anita, 262–63, 264, 265
Langeler, Ralf, 103
lanternfish, 35, 40, 42, 61; glacier, 36; Hector's, 42
Latham, John, 149, 152
leatherback sea turtles, 58, 63
Lemmens, Stijn, 233
lithium, 102
Lomonosov Ridge, 177–78, 180
London Protocol, 236
longlines, 63, 64, 76
Longyearbyen (Spitsbergen Island), 161–62
Lost City, 2, 117, 304, 306
lugworm, 96
Lukatela, Hrvoje, 228–29
lumpfish, 49–50
Lynch, Heather, 187, 201

Macau, 244
MacKay, Peter, 181
mackerel, 35, 36–37, 39, 43, 45, 62, 81, 311
Macron, Emmanuel, 125
mahi-mahi (common dolphinfish), 81, 82
Malaspina 2010 expedition, 65
manganese (polymetallic) nodules, 92–93, 95, 97–98, 105–6
Mare Liberum (Grotius), 16–17, 19–20
Mariana Trench, 13–14, 60
marine biomass regeneration, 141
marine biotechnology industry, 28, 211
marine cloud brightening (solar albedo modification), 146–47, 148, 149–51, 152–53
Marine Conservation Institute, 311
marine protected areas (MPAS): about, 273–74; Cabo Pulmo, 281–82; commercial activity within, 280; criteria for effectiveness, 280–81, 295; High Seas Treaty and, 30–31, 302, 303; implementation challenges, 274–75, 282–83; in Mediterranean, 295–96; Ross Sea, 201–2, 275, 281. *See also* Pelagos Sanctuary
marine snow, 51–54, 56–57, 58, 59–60, 141
Marine Stewardship Council, 46
MarinTrust, 45–46
Marizomib (salinosporamide A), 208, 227
Matzner, Eric, 136–37, 144
McKinnel, Tim, 72
McNutt, Marcia, 157

medicines, 27–28, 205, 207, 208, 209–11, 217, 221–22, 227
Medimetriks, 213, 217
Mediterranean Sea, 21, 278–79, 295–96. *See also* Pelagos Sanctuary
megafauna, 95
Melle, Webjørn, 61, 66, 68
Mendoza, Martha, 72
mercury, 237
Mero, John L., 100
mesopelagics, 35. *See also* twilight zone
The Metals Company, 114–15, 117, 118–19, 120–21
microbes, 205–7, 208–9, 214–16
microplastics, 243, 249, 293. *See also* plastics
midnight zone, 59–60
mining, deep-sea: alternatives to, 113; arguments for, 94, 101–2, 109, 112–13, 125; challenges facing, 94, 100–101; cobalt-rich crusts, 94, 95; environmental destruction from, 27, 93–94, 105, 106, 108–9, 123–24; environmental impact research, 102–8, 109–12, 113–15; green veneer, 102, 109; High Seas Treaty and, 304; manganese nodules, 92–93, 95, 97–98, 105–6; The Metals Company and, 114–15, 117, 118–19, 120–21; moratorium proposed, 124–25; Nautilus Minerals and, 119–20; profit-sharing proposal, 122–23; regulatory issues, 23, 24, 94, 115–18, 308; vs. seabed as common property, 98–100; seafloor massive sulphides (SMS), 94–95
mining, urban, 113
Mir (Soviet space station), 228, 230–32, 234
Mission Blue, 274
Mohammed bin Faisal Al Saud, 260
Mola mola (ocean sunfish), 297
Monaco, 273, 296
Mongolia, 71, 87
Monsanto, 219
Mooney, Pat Roy, 218–19, 220–21
Mougin, Georges, 260
Murray, John, 52
Musk, Elon, 135, 143, 230
MV *NIKA*, 89–90

Nagoya Protocol, 219, 223, 226
Nansen, Fridtjof, 166
National Academies of Science, Engineering, and Medicine (NASEM), 138–42, 144, 156–57, 235–36
National Cancer Institute, 223
Natural History Museum (London): Darwin Centre, 95
Nauru, 117, 118, 121–22
Nautilus Minerals, 119–20
neopelagic, 252
Netherlands, 14–16, 20, 235
Newfoundland, 257–59, 263, 267, 268–69
Newman, David, 223
New Zealand, 235
nickel, 95, 101, 118
Nicoud, Laurence, 199
Nike trainers, 245–46
noise pollution, 93, 279, 288–89
Nootroo (nootropics), 136
North Atlantic Garbage Patch, 2
North Korea, 184
North Pacific Gyre, 250
Norway: Arctic 5 agreement and, 181; Arctic claims, 179; fiscal challenge facing, 37–38; krill fishery, 184; Lomonosov Ridge and, 178; mesopelagics fishery, 24, 34–35, 35–36, 39–40, 41, 43–44, 68; overfishing disputes, 36–37, 45; salmon farming, 38–39, 46–50; Southern Ocean sanctuary and, 202
nuclear waste, 234, 235, 236

oarfish, 4
observers, fisheries, 74–75
ocean: acidity, 130–31; artificial upwelling and downwelling, 141–42; author's fascination with, 5–7, 8; chain of consumption, 56–57, 59–61; dead zones, 26–27, 131, 242; euphotic zone, 56; flotsametrics, 245–47, 249, 251; genetic resources, 220–21, 226;

gyres, boundary currents, and eddies, 250–51; hadal zone, 60; midnight zone, 59–60; Oxygen Minimum Zone (OMZ), 58–59; scattered productivity, 56; thermohaline circulation (global conveyor belt), 248–49. *See also* Arctic; biodiscovery; climate interventions; fisheries; high seas; icebergs; illegal, unreported and unregulated (IUU) fishing; marine protected areas; mining, deep-sea; pollution and ocean dumping; Southern Ocean; twilight zone
The Ocean Cleanup, 244–45
ocean dumping. *See* pollution and ocean dumping
ocean fertilisation, 126–29, 131–32, 139, 144–45
ocean genome, 225–26. *See also* genetic resources
Oceankind, 138, 144
ocean sunfish (*Mola mola*), 297
Ocean Surface Current Simulator (OSCURS), 246, 247
octopus, 82, 93, 106, 131; dumbo, 60, 108
oil: Russian production plans in Arctic, 170; spills, 75, 160, 238–40, 313
Ok Tedi (mining disaster), 121
Oman, 42
open registers (flags of convenience), 86–87, 160, 308
optical imagery, 88
orange roughy, 7, 8–9
Orheim, Olav, 260
overfishing. *See* fisheries; illegal, unreported and unregulated (IUU) fishing
Oxygen Minimum Zone (OMZ), 58–59

Pacific Fisheries Management Council, 43
Pacific Ocean, 9, 56, 57
Panama, 85–87, 88–89
Panama Papers, 86
Papua New Guinea, 119, 121, 124–25
PARC (Palo Alto Research Center), 152
Pardo, Arvid, 98–100
Paris Climate Accord, 132, 305, 311, 313
Particularly Sensitive Sea Area (PSSA), 296
passenger pigeon, 201
Patania II (deep-sea mining machine), 114
Patent and Trademark Office (US), 219
pearlsides, 35, 42–43, 61; Mueller's, 36
pederin, 214
pelagic waters, 5. *See also* high seas
Pelagos (Tethys boat), 276, 277–78
Pelagos Sanctuary: about, 272–73; author's experience, 283–84, 285–86, 286–88, 290–91; as failed marine sanctuary, 275, 279–80, 295; high seas and, 289–90; plastic in, 292, 293, 294; strengthening protections, 296–98; whales, 283–84, 285–86, 286–88, 290–91
Penguin Island, 200–201, 202
penguins: Adélie, 185, 186, 191, 202; chinstrap, 185–86, 191–94, 196, 198, 199, 200–201; emperor, 191, 202; gentoo, 191, 192, 198, 204; king, 184, 204; Magellanic, 188
Peru, 40, 80, 81, 83–84
Peru Basin, 110, 111
Peruvian anchovy (anchoveta), 40, 81–82
pharmaceutical industry. *See* medicines
PharmaMar, 213, 217
Phoenicians, 278
phosphate mining, 121–22
phytoplankton, 51, 56, 57, 58, 267. *See also* plankton
Pierantonio, Nino, 273, 276–77, 283–84, 285, 287, 289, 291, 294, 295, 297–98
pigeon, passenger, 201
piracy, 10. *See also* biopiracy; illegal, unreported and unregulated (IUU) fishing
Pisanu, Francesco 'Frankie', 168–70
Pitcairn Islands, 229, 282
Planetary Technologies, 143
plankton: in Arctic, 175–76; bilge water and, 241; in Costa Rica Thermal Dome, 57–58; marine biomass regeneration and, 141; marine snow and, 56–57; migration to surface, 51, 113; ocean

fertilisation and, 126–29, 139; plastic and, 293. *See also* phytoplankton; zooplankton
plastics, 27, 242–45, 249, 279, 292–95, 304–5
Point Nemo, 228–29, 243
Poland, 117
Polar Code, 160
Polar Silk Road, 158–59
pollution and ocean dumping: animals hitching rides on debris, 251–52; bilge water, 240–41; challenges tackling, 252–53; currents and, 245–48, 249, 251; High Seas Treaty and, 304; history of, 235–36; international regulations, 236, 239; legacy pollutants, 237–38; noise, 93, 279, 288–89; nuclear waste, 234, 235, 236; oil spills, 75, 160, 238–40, 313; as ongoing problem, 26–27, 237; in Pelagos Sanctuary, 279, 293; plastics, 27, 242–45, 249, 279, 292–95, 304–5; space waste, 26, 228–35
polychaetes, 96, 97
Pompeii worm, 223
Porcupine Abyssal Plain, 60
Port State Measures Agreement, 305
Portugal, 14–16, 20
positioning, 43–44
Prakoso, Imam, 90
Pratt, E.J., 265
Prialt, 210
Prince Albert II of Monaco Foundation, 293–94
Project Vesta, 137, 144
protected areas. *See* marine protected areas
pupukeanane, 213
Pusaka Benjina Resources, 72
Putin, Vladimir, 170, 179, 181

Qureshi, Jamal, 264

rabbitfish, 7
Rackete, Carola, 199–200
Rau, Greg, 132, 134, 136, 142, 143, 157
rays, 63
recycling, 113
Red Sea, 110
regional fisheries management organisations (RFMOS), 23, 62–63, 75, 304
Remdesivir, 28, 210, 221–22
Renna, Angelo, 292
res communis (common property), 98–99
res nullius, 98
ribbonfish, 59
Ricketts, Ed, 281–82
Riebesell, Ulf, 135, 141–42
Rio Group, 232, 234
Rios, Jorge, 70, 73
Rio Tinto, 121
Romans, ancient, 98
rooibos, 219
Ross Sea marine protected area, 201–2, 275, 281
rove beetle, 214
Royal IHC, 104, 105–6
Russell, Colin, 169–70
Russia: Arctic 5 agreement and, 181; Arctic 30 protesters and, 169–70; Arctic claims, 25, 178–79, 180–81; Arctic shipping regulations, 160; continental shelf, 22; on disposing space waste at sea, 232, 234; krill fishery, 185; Lomonosov Ridge and, 177; Northeast Passage, 159; overfishing, 36; twilight zone fishery, 42. *See also* Soviet Union

sacrifice zones, 108
Sala, Enric, 311
Salinispora, 208
salinosporamide A (Marizomib), 208, 227
salmon, farmed, 38–39, 46–50
salps, 52, 55, 199
Salter, Stephen, 149, 152
Salwai, Charlot, 124
Sanchi oil spill, 239–40
sanctuaries. *See* marine protected areas
Sanremo (Italy), 272, 275–76
Santa Catarina (Portuguese merchant ship), 14–16
Sargasso Sea, 2, 215, 306
Sarmiento de Gamboa (research vessel), 102–4, 106–7, 125
satellites, 230
Saudi Arabia, 260, 309

Saya de Malha Bank, 2–3
scaly-foot snail, 225, 306
Scheuer, Paul, 211–13
Schneider, Stephen, 146
Schriever, Gerd, 111
Scripps Institution of Oceanography, 207
seabed, 98–100, 102, 179–80. *See also* mining, deep-sea
Seabed 2030 project, 4
sea butterflies, 106
sea cucumbers, 95, 281
seafloor massive sulphides (SMS), 94–95
sea ice: climate change and, 25–26, 161, 162, 167, 171–72, 174; restoration efforts, 153–56. *See also* icebergs
sea lice, 49–50
sea lilies, 12
seals: crabeater, 185; elephant, 183–84, 195, 200; fur, 183–84, 200, 203; Weddell, 200
sea mouse, 96
Sea Shepherd, 71, 77–78, 89
sea slugs, 217; fried egg, 212–13
sea snails, 210, 281, 313; scaly-foot, 225, 306; violet (bubble raft), 251
sea stars (starfish), 95, 281
SeaTech Solutions, 120
sea urchins, 281
seaweed cultivation, 139–40
Senkaku Islands, 108
Shackleton, Ernest, 194–95
sharks, 58, 59, 61, 63, 64, 67, 77, 78–79, 313; great white, 57; hammerhead, 78, 180; whale, 78
shrimp: eyeless, 2; giant red mysid, 59
SilverLining, 145–46
SINTEF SeaLab, 35–36, 37–38, 40, 43, 65–66
siphonophores, 32, 33, 55
SkyTruth, 73, 75, 84–85, 87, 241. *See also* Global Fishing Watch
slavery, at sea, 71, 84, 85, 308–9
Sloane, Nick, 261–62
Sloane's viperfish, 33
slugs. *See* sea slugs
Smith, William, 183
Smithsonian Environmental Research Center, 252
snails. *See* sea snails
Sneider, David, 137
snow, marine, 51–54, 56–57, 58, 59–60, 141
solar albedo modification (marine cloud brightening), 146–47, 148, 149–51, 152–53
solubility pump, 54
Solwara 1 (deep-sea mining project), 119, 121, 124
Somali Current, 250
Sosik, Heidi, 54–55
soundscape, ocean, 288–89. *See also* noise pollution
South Africa, 42, 218, 219, 261–62
Southern Ocean: Cape Horn and Drake Passage, 188–89; changes in, 202–4; chinstrap penguins, 185–86, 191–94, 196, 198, 199, 200–201; climate change and, 185, 198; fuel ban, 160–61; governance of, 196–97; Greenpeace expedition, 186, 187–88, 189–91, 192–96, 199–201, 203–4; historic exploitation, 183–84; krill, 25–26, 38, 184–85, 192, 196–99, 202; marine protected areas, 201–2; ocean fertilisation trial, 139; pirate fishing in, 71; Shackleton's expedition, 194–95; thermohaline circulation and, 249; whales' return to, 313
South Korea, 185, 202, 235, 244
South Orkney Islands, 274
South Shetland Islands, 183, 197, 202, 203. *See also* Penguin Island
Soviet Union (USSR): cloud seeding and, 151; krill fishery, 184; Mir space station, 228, 230–32, 234; northern fisheries, 20; North Pole expedition, 166; ocean dumping, 235, 236. *See also* Russia
space: debris landing outside Point Nemo, 232–33; human-made objects circling Earth, 229–30; International Space Station (ISS), 229, 232; marine impacts from waste, 233–35; Mir space station, 228, 230–32, 234; Point Nemo spacecraft cemetery, 228–29, 230, 232; satellites, 230
SpaceX, 230
Spain, 43, 65, 129, 301

Spanish National Research Council (CSIC), 104
sperm whales: Antarctic whaling and, 184; diving, 290; identifying by whale fluke, 291; ocean fertilisation trial and, 128; in Pelagos Sanctuary, 273; Pierantonio on, 276–77; plastic and, 293, 295; ship strikes, 279, 291–92; surfacing, 286–88, 295; vocalisations, 283, 284–86, 289
SpeSeas, 96
squid: in Costa Rica Thermal Dome, 58; fisheries and pirate fishing, 26, 64, 74, 80–81, 82–83, 87–88, 90–91; ocean fertilisation trial and, 128
squid, specific species: Argentine shortfin, 80–81, 83, 91; Japanese flying, 83; jumbo (Humboldt), 80–81, 91; neon flying, 83; vampire, 59–60
Standal, Dag, 37–38, 43, 67
starfish (sea stars), 95, 281
Steinbeck, John, 281–82
Stockholm Convention, 236
Stockholm Resilience Centre, 28
Straits of Magellan, 188–89
Streptomyces, 207
Strycker, Noah, 186–87, 188, 189, 191, 193, 194, 196, 199, 204
Suez Canal, 160
Sumaila, Rashid, 311
Superflex: *Vertical Migration* installation, 32, 33
Sustainable Development Goals, 264
Svalbarði, 264
Sweden, 165, 178, 235
Sweep Island, 292
Swire, Herbert, 13–14
Switzerland, 235
swordfish, 311
synthetic aperture radar (SAR), 88

Taiwan, 244
Tarling, Geraint, 266–67, 270
Tethys Research Institute, 273, 275, 277
thermohaline circulation (global conveyor belt), 248–49
Thiel, Hjalmar, 109, 110–11
Thomson, Charles Wyville, 11–12, 52, 53, 97–98
thorium, 53
time-series, 167
Tonga, 117, 121, 122
toothfish, 71, 73, 184, 199; Patagonian, 64, 197
Torrey Canyon oil spill, 238–39, 241
tragedy of the commons, 26–27
transhipment, 85, 90
Transpolar Drift, 171–72
Trathan, Phil, 198
trawling, bottom, 8–9, 20–21, 29, 64, 93, 279–80, 300–301, 302
tributyltin (TBT), 313
tuna: in Arabian Sea, 59; in chain of consumption, 57; in Costa Rica Thermal Dome, 58; deep-sea mining and, 106; drift nets and, 77; management and conservation, 62, 63, 313; near San Diego, 207; ocean fertilisation and, 128; overfishing, 61, 63; pirate fishing and, 71, 73; profits from, 64; as transboundary stocks, 311
tuna, specific species: albacore, 63; bluefin, 64, 278; Pacific bluefin, 26, 63; skipjack, 63; yellowfin, 39, 63
turtles, 57, 58, 63, 77, 313
twilight zone: about, 24, 32–33; aquaculture fish food and, 49–50; carbon storage in, 24, 33–34, 39, 50–51, 53–54; exploration of, 54–56; fishing in, 24, 34–35, 42–43, 61–62, 65–68; marine snow, 51–54, 57, 59–60; Norwegian mesopelagics fishery, 24, 34–35, 35–36, 39–40, 41, 43–44, 68; odd inhabitants, 33

UNESCO, 58, 282
Union of Concerned Scientists, 230
United Arab Emirates, 254–55, 256, 260–61, 262
United Kingdom: Antarctic exploitation, 183–84; Cod Wars, 20–21; flotsam arriving in, 247; Grotius' *Mare Liberum* and, 19–20; iron fertilisation and, 129; marine genetic resources and, 220, 221; marine protected areas and, 282; nuclear dumping, 235; overfishing and, 36, 43

United Nations: Convention on Biological Diversity (CBD), 129, 274; Convention on the Law of the Sea (UNCLOS), 21, 22, 23, 219–20, 226; Fish Stocks Agreement, 62; Framework Convention on Climate Change, 305; on icebergs, 262; on seabed, 98–100, 102, 179–80; Sustainable Development Goals, 264
United States of America: Antarctic exploitation, 183–84; Arctic and, 178, 181; Clean Water Act, 236; cloud seeding, 151; Convention on the Law of the Sea and, 23; marine genetic resources and, 220, 221, 226; mesopelagic fishery moratorium, 43; nuclear waste dumping, 235; offshore fishing fleet, 83; plastics from, 244
University of California, Davis, 219
urban mining, 113
USSR. *See* Soviet Union

van Heemskerck, Jacob, 14–15
Vanuatu, 124
Velella velella (by-the-wind sailor), 251
Venter, Craig, 214–15, 216, 219
Venus flower basket, 12–13
vessel monitoring system (VMS), 85–86, 88, 90
Vilm (Baltic Sea island), 299–300
violet sea snail (bubble raft snail), 251
Visible Infrared Imaging Radiometer Suite (VIIRS), 87–88
Vulnerable Marine Ecosystems (VMES), 274

Wadhams, Peter, 260, 267–68
Wagner, Till, 163, 167, 172, 174
Wanser, Kelly, 145–47, 150–51, 152, 156, 157
water, drinking, from icebergs, 254–59, 260–65
weather modification, 151
Weddell Sea, 194, 202, 249
whales: in Antarctic, 184, 313; in Arabian Sea, 59; in Arctic, 175; carbon storage and, 140–41; early whaling, 11; identifying by whale flukes, 291; overfishing, 61; in Pelagos Sanctuary, 273, 279, 286–88, 290–92; plastics and, 293, 294–95; ship strikes, 279, 291–92; vocalisations, 283–86, 287–88, 289
whales, specific species: beluga, 175; blue, 57, 58, 184, 207; bowhead, 175; Cuvier's beaked, 273; fin, 128, 184, 273, 279, 284, 290–91, 293, 294, 313; humpback, 184, 198, 203, 284, 313; minke, 184; narwhals, 175; orcas, 128, 202; pygmy blue, 3; right, 184; sei, 128, 184; sperm, 128, 184, 273, 276–77, 279, 283, 284–86, 286–88, 289, 290–91, 293, 295
White Shark Café, 57, 58, 304, 306
Wilson, E.O., 312
Winderen, Jana, 288
Woods, Paul, 75, 87
World Bank Group, 101
World Economic Forum, 262
World Trade Organization (WTO), 305, 309
World Wildlife Fund, 296
wrasse, 49–50

Xuelong 2 (Chinese icebreaker), 158, 159

Yondelis, 210
Yongle (Chinese emperor), 9–10

Zheng He, 9–10
zinc, 95, 178
zooplankton, 40, 51, 53, 57, 127, 141, 175, 176, 241. *See also* plankton

DAVID SUZUKI INSTITUTE

THE DAVID SUZUKI INSTITUTE is a companion organization to the David Suzuki Foundation, with a focus on promoting and publishing on important environmental issues in partnership with Greystone Books.

We invite you to support the activities of the Institute. For more information, please contact us at:

David Suzuki Institute
219 – 2211 West 4th Avenue
Vancouver, BC, Canada V6K 4S2
info@davidsuzukiinstitute.org
604-742-2899
davidsuzukiinstitute.org

Cheques can be made payable to The David Suzuki Institute.